WHAT DOES A

Black Hole Look Like?

WHAT DOES A

Black Hole Look Like?

CHARLES D. BAILYN

PRINCETON UNIVERSITY PRESS

PRINCETON AND OXFORD

Library of Congress Cataloging-in-Publication Data

Bailyn, Charles D., author.
    What does a black hole look like? / Charles D. Bailyn.
        pages cm. – (Princeton frontiers in physics)
    Includes index.

    Summary: "Emitting no radiation or any other kind of information, black
holes mark the edge of the universe–both physically and in our scientific
understanding. Yet astronomers have found clear evidence for the existence of
black holes, employing the same tools and techniques used to explore other
celestial objects. In this sophisticated introduction, leading astronomer Charles
Bailyn goes behind the theory and physics of black holes to describe how
astronomers are observing these enigmatic objects and developing a remarkably
detailed picture of what they look like and how they interact with their
surroundings. Accessible to undergraduates and others with some knowledge of
introductory college-level physics, this book presents the techniques used to
identify and measure the mass and spin of celestial black holes. These key
measurements demonstrate the existence of two kinds of black holes, those with
masses a few times that of a typical star, and those with masses comparable to
whole galaxies–supermassive black holes. The book provides a detailed account
of the nature, formation, and growth of both kinds of black holes. The book
also describes the possibility of observing theoretically predicted phenomena
such as gravitational waves, wormholes, and Hawking radiation. A cutting-edge
introduction to a subject that was once on the border between physics and
science fiction, this book shows how black holes are becoming routine objects
of empirical scientific study."– Provided by publisher.
    ISBN 978-0-691-14882-3 (hardback : acid-free paper) – ISBN 0-691-14882-1
(hardcover : acid-free paper) 1. Black holes (Astronomy) 2. Astrophysics. I. Title.
    QB843.B55B35 2014
    523.8′875–dc23

                        2014009784

British Library Cataloging-in-Publication Data is available

This book has been composed in Garamond and Helvetica Neue

Printed on acid-free paper. ∞

Typeset by S R Nova Pvt Ltd, Bangalore, India

Printed in the United States of America

10 9 8 7 6 5 4 3 2 1

For D.

CONTENTS

PREFACE

The goal of this book is to introduce readers to the empirical study of black holes. Readers are assumed to have some knowledge of basic college-level physics and mathematics. The focus is on current understanding and research relating to astrophysical manifestations of black holes rather than on the underlying physical theories. Nevertheless, an understanding of the physics is necessary to understand and interpret observations. So, I have presented descriptions and derivations of the physical processes in a way that I hope will illuminate the observations and have focused on the physical principles involved rather than the full presentation needed to solve detailed problems. Interested readers should consult the many excellent textbooks in the field for further discussions. In particular, Bernard Schutz's book *A First Course on Relativity*; Rybicki and Lightman's *Radiative Processes in Astrophysics*; Shapiro and Teukolsky's *Black Holes, White Dwarfs and Neutron Stars: The Physics of Compact Objects*; and Frank, King, and Raine's *Accretion Power in Astrophysics* are all classic texts at the upper undergraduate and introductory graduate student level that present the relevant physics clearly and in detail. The focus in this text is on the observational astrophysics—on what is observed and what can be inferred from the observations—in short, on what black holes look like.

I am very fortunate to have lived my life as part of a community of incisive thinkers who have helped me learn and understand the material in this book, and many other things besides. In particular, I am grateful to my undergraduate advisors-turned-colleagues Pierre Demarque, Bob Zinn, and Richard Larson, who have been teaching me and encouraging me for more years than any of us would care to admit; to my graduate advisors Peter Eggleton and Josh Grindlay, who introduced me to the wonderful world of interacting binary stars; to my long-time collaborators Jeff McClintock, Ron Remillard, and Jerry Orosz, who have worked with me for two decades to ferret out the characteristics of stellar-mass black holes; to my many friends and colleagues who have made Yale such a wonderful place to work on high-energy astrophysics, including but not limited to Meg Urry, Paolo Coppi, Andy Szymkowiak, Michelle Buxton, Ritaban Chatterjee, and Erin Bonning; and to my outstanding students-turned-colleagues who have struggled with me to understand the many manifestations of black holes, particularly Raj Jain, Dipankar Maitra, Andy Cantrell, Laura Kreidberg, Jedidah Isler, and Rachel MacDonald. Finally, I must express gratitude and love for my learned and loving nuclear and extended family, most particularly to the gentleman to whom this book is dedicated, who taught me more about how to think than anyone else. That I have anything at all to say on this or any other scientific topic is due largely to conversations with these and many other interlocutors, both in person and in the published literature. But any deficiencies in understanding or in exposition are mine alone.

WHAT DOES A

Black Hole Look Like?

1

INTRODUCING BLACK HOLES: EVENT HORIZONS AND SINGULARITIES

Black holes are extraordinary objects. They exert an attractive force that nothing can withstand; they stop time, turn space inside out, and constitute a point of no return beyond which our universe comes to an end. They address issues that have always fascinated humans—literature and philosophy in all times and cultures explore irresistible lures, the limits of the universe, and the nature of time and space. In our own time and place, science has become a dominant force both intellectually and technologically, and the scientific manifestation of these ancient themes provides a powerful metaphor that has come to permeate our culture—black holes abound not only in speculative fiction but in discussions of politics, culture, and finance, and in descriptions of our internal and public lives.

But black holes are not just useful metaphors or remarkable constructs of theoretical physics; they actually exist. Over the past few decades, black holes have moved from theoretical exotica to a well-known and carefully studied class of astronomical objects. Extensive data archives reveal the properties of systems containing black holes, and many details of their behavior are known. In the current astronomical literature, the seemingly bizarre properties of black holes are now taken for granted and are used as a basis for understanding a wide variety of phenomena.

The title of this book is oxymoronic. The defining property of black holes is that they do not emit radiation (hence "black")—so they cannot "look like" anything at all. Nevertheless, black holes are the targets of a wide variety of observational studies. This paradox is of a piece with much of modern astrophysics, in which objects that cannot be observed directly are studied in detail. Cosmologists have found that more than 90% of the mass energy of the Universe is in the form of unobservable "dark matter" and "dark energy." Thousands of planets have been discovered orbiting stars other than our Sun, but only a tiny handful have been observed directly. So it is with black holes. They cannot be observed directly, and yet they can be studied empirically, in some detail.

My goal for this book is to describe how astronomers carry out empirical studies of a class of objects that is intrinsically unobservable, and what we have found out about them. I will focus on current observations and understanding of the astrophysical manifestations of black holes, rather than on the underlying physical theories. There are a number of excellent textbooks on the physical processes, and I will refer to them along the way. The first three chapters sketch some of the physics needed to understand and interpret the observational results. Subsequent chapters describe observed black holes, and thus provide an answer to the question, What do black holes look like?

1.1 Escape Velocity and Event Horizons

One of the basic concepts to emerge from Newton's theory of gravity is the *escape velocity*, denoted V_{esc}. The escape velocity is the speed required to escape the gravitational attraction of a spherical object. It can be shown from basic principles that

$$V_{esc} = \sqrt{2GM/R}$$

where G is the gravitational constant (equal to $6.674 \times 10^{-11} \, \mathrm{m^3 \, kg^{-1} \, s^{-2}}$), and M and R are the mass and radius, respectively, of a spherical object.[1]

It is a simple matter to calculate the escape velocity for any combination of size and mass. For example, numbers approximating the size and mass of a human being (1 m and 50 kg) result in an escape velocity of just over $80 \, \mu\mathrm{m \, s^{-1}}$ (or about a foot per hour). While this result would apply precisely only to a spherical object with $R = 1$ m and $M = 50$ kg, an object of comparable mass and size would have a comparable escape velocity. Because the resulting escape velocity is much slower than the speeds associated with everyday life, gravitational effects between ordinary objects (people, cars, buildings) can generally be ignored. By contrast, the Earth, with a mean radius of a bit less than 6400 km and a mass of 5.9×10^{24} kg, has an escape velocity of $11 \, \mathrm{km \, s^{-1}}$—much faster than everyday speeds. So, without mechanical assistance we remain bound to the Earth.

[1] Technically, the escape velocity thus calculated applies to test particles attempting to escape from the surface of the sphere.

The most conceptually straightforward description of a black hole is an object whose escape velocity is equal to or greater than the speed of light. Such objects had already been contemplated in the eighteenth century.[2] In such a case, we can rewrite the escape velocity equation as

$$R \leq R_s = 2GM/c^2,$$

where c is the speed of light, and R_s is the *Schwarzschild radius* (named after the early twentieth-century physicist Karl Schwarzschild). In the context of Newtonian physics such objects have no particularly striking physical qualities other than their small size (R_s of a mass equal to that of the Earth is only about a centimeter). Presumably, light would not be able to escape from them, so they would be hard to observe. But the fascinating physics associated with black holes emerged only when *general relativity* was developed.

Nevertheless, it is amusing to play with the Newtonian concept of black holes as objects with $V_{esc} \geq c$ and to notice how the size and density of black holes vary with their mass. The density ρ of an object is defined as mass/volume, so the density of a black hole must be

$$\rho_{bh} \geq \frac{3}{32\pi} \frac{c^6}{M^2 G^3}.$$

Thus the density required to form a black hole decreases as the mass of the black hole increases—masses 10^8 times that of the Sun (which, as we will see, are common in

[2]The eighteenth century British philosopher John Michell is generally credited with the first published consideration of objects with escape velocities greater than c. See Gary Gibbon 1979, "The Man Who Invented Black Holes," *New Scientist* 28 (June) 1101.

the center of large galaxies) will become black holes even if their density is no greater than that of water. Black holes with masses comparable to those of typical stars must attain densities comparable to those of atomic nuclei, that is, much greater than the density of ordinary matter. Less massive black holes would require densities far beyond that of any known substance.

In the general theory of relativity, the Schwarzschild radius becomes fundamentally important. Gravity is not considered a force in general relativity but, rather, is a consequence of the curvature of space-time. Mass causes space-time to curve, and this curvature affects the trajectories of objects. In situations where the distances between objects are large compared with their Schwarzschild radii, the predictions of general relativity become indistinguishable from those of Newtonian gravity, and all the familiar Newtonian results can be recovered. However, as distances between objects approach R_s, objects begin to behave differently from Newtonian predictions. Indeed, the first observational evidence supporting general relativity came from slight anomalies in the orbit of Mercury, the planet closest to the Sun. Mercury's mean distance to the Sun is about 20 million times the Schwarzschild radius associated with the mass of the Sun, so the deviations are quite small, but the orbits of the planets are known very precisely, so the deviation was already known before Einstein developed his theory. Closer to the Schwarzschild radius, the differences between Newtonian and relativistic physics become greater, leading eventually to drastic qualitative differences in behavior. These dramatic effects cannot be observed in Earth-bound laboratories, or indeed anywhere in the solar system, because all nearby objects have radii that are many orders of magnitude bigger than R_s. But as we will see, black holes and the dramatic physical effects associated with them can be found in other astronomical contexts.

For objects that fit inside their Schwarzschild radius, the spherical surface where $r = R_s$ is often referred to as the *event horizon*. This name comes about because information from inside the event horizon cannot propagate to the outside world. Consequences of events that occur at $r < R_s$ cannot be seen by an observer outside R_s. The interior of the event horizon is thus causally disconnected from the rest of the Universe—in a sense, it is not part of our Universe. The behavior of matter and energy inside the event horizon can be explored mathematically by assuming that the equations of general relativity apply and then interpreting the results of mathematical manipulations of these equations. However, the laws that lead to the equations also categorically prohibit them from being tested by experiments or observations conducted by observers located at $r > R_s$. Thus from an epistemological point of view, physics inside an event horizon is a different kind of science from physics in parts of the Universe that are causally connected to us.

1.2 The Metric

We will not explore the details of the mathematics associated with general relativity here.[3] But simply looking at the

---

[3]For a good introduction at the undergraduate level, see Bernard Schutz, *A First Course in General Relativity* (Cambridge: Cambridge University Press, 2009).

form of some of the relevant equations can reveal some of the remarkable qualities of black holes.

Mathematically, the curvature of space-time is defined by a *metric*. A metric defines a line element ds, and the separation between two space-time events is given by the integral of ds. In general, this integral depends on the trajectory taken by the object. In the absence of external forces, objects follow trajectories that minimize the separation.[4] In an uncurved space-time, this behavior is in accordance with Newton's first law of motion, which requires that (in the absence of forces) objects move in straight lines (the closest distance between two points). In relativity, objects in a gravitational field follow a curved trajectory in space not because of a "gravitational force" that redirects their motion but rather because of the curvature of space-time itself: the minimum separation between two space-time events follows a curve in spatial coordinates.

A single point mass generates a space-time curvature associated with the so-called Schwarzschild metric:

$$ds^2 = -(1 - R_s/r)c^2 dt^2 + \frac{dr^2}{1 - R_s/r} + r^2 d\Omega^2.$$

Here space is measured in polar coordinates (r, Ω, where $d\Omega = \sin\theta\, d\theta\, d\phi$), with the point mass at the origin. To be specific, dt is the time interval seen at infinity, and R_s is the circumference around the black hole divided by 2π. By looking at the limiting cases of this equation, we can gain some insight into how black holes behave.

---

[4] Formally, the "proper time" is maximized.

$r \rightarrow \infty$: When r is large, the gravitational influence is small. In this case the spatial coordinates of the metric approximate polar coordinates of a Euclidean space, and time and space decouple. In this limit, the equations of general relativity reduce to the familiar results of Newtonian physics.

$r \rightarrow R_s$: As r approaches the Schwarzschild radius, the term $(1 - R_s/r)$ goes to zero. The time term of the metric thus becomes zero, and the radial term becomes infinite. Something very peculiar must happen at $r = R_s$! Indeed, as an object falls toward the black hole, it appears to an outside observer that time is slowing down. That is, a clock mounted on the infalling object runs slower than an identical clock that remains at a large distance from the black hole. This observed slowness applies also to the frequencies of emitted radiation. Radiation emitted near a black hole will be observed to have lower frequencies, and thus longer wavelengths. This effect is called *gravitational redshift* and can be observed in a variety of ways (see chapter 8). However, for an observer on the infalling object, local clocks appear to be accurate, the Universe far from the black hole appears to speed up, and the radiation from distant objects appears to be blueshifted.

$r < R_s$: At radii smaller than R_s, the term $(1 - R_s/r)$ becomes negative. This means the signs of the time and radial terms of the metric are reversed. As a consequence, radial motion can be only unidirectional (as time is in ordinary situations), while it is possible, in principle, to move forward and backward in time. Inside the Schwarzschild radius "time machines" are thus in principle possible. This property of black holes has led to a wide variety of speculative fiction, and some interesting physics as well. But inside the event horizon, it is not possible to move outward, any more than it is possible to move backward in time in less exotic regions of space-time. Any object that finds itself inside the Schwarzschild radius of a black hole will inexorably travel toward the center of the coordinate system at $r = 0$. Thus material will pile up in a point of zero volume and infinite density at the center of a black hole. This point is sometimes referred to as a *singularity*.

There are thus a number of situations related to black holes in which physical quantities should become infinite. At $r = R_s$, terms of the metric become infinite, and time stops.[5] At the center of the black hole, where $r = 0$, the density of matter becomes infinite. The existence of this central singularity suggests that the physical theory is likely to be incomplete.[6] In particular, there are likely to be quantum effects (which become important at small sizes) that need to be accounted for. But relativity is a continuous theory and does not fit easily with quantum mechanics.

---

[5] The mathematical divergence of the terms of the metric at $R_s = r$ can be avoided by an appropriate change in coordinates, but that does not change the predicted behavior of an infalling object as observed from a long distance away.

[6] An example is the "ultraviolet catastrophe'" in radiation theory, in which classical physics requires the radiation emitted from a blackbody to become infinite at high frequencies. One of the first triumphs of quantum physics was to eliminate this infinity from the theory.

The search for a "unified theory" which combines general relativity and quantum mechanics is ongoing. Until this search is complete, physical predictions of the behavior of matter and energy at the Schwarzschild radius or near the central singularity might plausibly be regarded as the results of an incomplete theory rather than as any sort of accurate representation of reality. One of the long-term goals of observational relativistic astrophysics is to probe this regime where current theory might encounter difficulties. However, as we will see, there are no current observations that contradict general relativity.

The Schwarzschild metric is actually a special case that applies only to black holes with no angular momentum. If the material forming the black hole has angular momentum (as one would expect to find in any physical object, particularly if it forms from the collapse of a much larger structure), then a more complicated metric known as the *Kerr metric* is the correct description of space-time. The Kerr metric has a key parameter in addition to the mass of the central object, namely, its angular momentum, usually given in dimensionless units as $a = J/(GM^2/c)$, where J is the angular momentum of the object. If $a \leq 1$, the Kerr metric generates an event horizon at

$$R_K = (GM/c^2)(1 + \sqrt{1 - a^2})$$

(note that this reduces to the Schwarzschild radius when $a = 0$). Situations in which event horizons exist and $a > 1$ are generally thought to be nonphysical. In Kerr black holes, the central singularity takes the form of a ring, rather than a point, and there is an additional critical surface

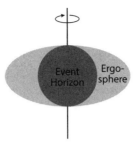

Figure 1.1. The event horizon and the ergosphere of a spinning black hole.

outside the event horizon called the *ergosphere*, inside of which objects cannot remain stationary but can escape from the black hole (see figure 1.1). Exchange of energy between particles within the ergosphere, some of which escape to infinity and some of which fall into the event horizon, allows much of the rotational energy of a Kerr black hole to be extracted and transferred to the outside universe.[7] The observable consequences of black hole spin will be explored further in chapter 8.

1.3 What Is a Black Hole?

So what exactly is a "black hole"? The term "black hole" is not defined in a technical way and is used in different contexts to mean different things. The phrase itself was popularized by the physicist John Archibald Wheeler to replace the cumbersome description "gravitationally

[7]R. Penrose, 1969, *Rivista del Nuovo Cimento 1* (Ser. 1): 252.

completely collapsed object."[8] The term can be used to describe an object whose escape velocity is greater than the speed of light, which leads to a quasi-Newtonian description of such objects. Sometimes "black hole" is used to denote the volume inside the event horizon, that is, the region "outside our Universe." In this context it makes sense to discuss the "size" or "density" of a black hole, since there is a nonzero radial distance (R_s in the case of a nonspinning black hole) associated with the object. Sometimes "black hole" is used to refer specifically to the singularity, in which case such physical quantities are not well defined. Finally, "black hole" has become a commonly used metaphor for anything with an inexorable pull leading to destruction. As we will see, the assumptions about black hole behavior associated with these metaphors are often quite misleading when applied to the physical objects themselves.

---

[8] By his own account, Wheeler himself did not invent the term. Rather, the phrase was called out by an anonymous voice at a conference, and Wheeler adopted it then and afterwards. J. A. Wheeler, *Geons, Black Holes, and Quantum Foam* (New York: Norton, 2000), 296–97.

2

ACCRETION ONTO A BLACK HOLE

One might think that it would be difficult to observe black holes, given that their defining characteristic is that they do not emit light. However, their presence is clearly detected through their gravitational effects on nearby objects. In particular, gas accreting onto a black hole generates huge amounts of energy that create easily observable effects. In fact, accretion energy powers the most luminous objects in the Universe and can be much more efficient at turning mass into energy than the thermonuclear processes that power ordinary stars like the Sun. The fusion of hydrogen atoms to make more massive nuclei generates energy equivalent to just under 1% of the mass of the hydrogen gas. But accretion onto a black hole can produce energy much more efficiently, depending on the way the gas flows onto the black hole and the way that the gas transforms kinetic energy into radiation. In this chapter we will explore the ways that accretion produces energy and radiation.

As material falls into a gravitational potential well, energy is transformed from gravitational potential energy into other forms of energy, so that total energy is conserved. If all the energy is turned into kinetic energy, as in the case of a single object in free fall, then

$$V^2/2 - GM/D = \text{constant},$$

where V is the velocity of the infalling object, M is the mass of the accretor, and D is the distance between the two. As the object falls in, D decreases, so V must increase. This kinetic energy can in turn be converted into other forms of energy, and thence into detectable radiation. In particular, an infalling stream of gas can convert kinetic energy of the individual gas particles into heat energy of the gas as a whole. Hot gas glows, and thus the infalling gas is a radiation source. The deeper the material falls into the potential well, the more energy it can, in principle, pick up. Black holes have the deepest possible potential wells, so accretion onto a black hole is especially energetic. Observing such accretion energy is one of the primary ways that astrophysicists pinpoint the locations of potential black holes. The spectrum and intensity of this radiation is governed by the geometry of the gas flow, the mass infall rate, and the mass of the accretor. Thus careful study of the accretion radiation, and comparison with detailed models of accretion, can reveal not only the presence of a black hole but the ways in which the infalling matter is influenced by the strong relativistic effects that the black hole creates.

2.1 Spherical Accretion and the Eddington Limit

The simplest flow geometry is that of a stationary object accreting mass equally from all directions. Such spherically symmetric accretion is referred to as *Bondi-Hoyle accretion*. As the gas falls inward it heats up and releases radiation. This radiation can be intercepted by gas farther out, which can retard the flow. Bondi-Hoyle accretion is thus self-limiting, in the sense that a large infall can generate enough radiation to halt the inflow altogether. This limit is called the *Eddington limit*, and it occurs when the gravitational force inward is balanced by the outward force generated by the radiation. In the case of ionized hydrogen gas, the gravitational force is dominated by the force on the protons, which is $F_{\text{in}} = GMm_p/r^2$, where M is the mass of the accreting object, and m_p is the mass of the proton. The gravitational force on the electrons is negligible, since the mass of an electron is so much smaller than the mass of a proton.

By contrast, the outward force generated by the radiation is dominated by the interaction of the radiation with the electrons. The force on a single electron can be computed by multiplying the radiation flux divided by the speed of light c by a cross section which is the effective geometric area that the electron presents to the radiation field. The radiation pressure is given by the intensity of the radiation field I divided by the speed of light c, and the relevant cross-sectional area is the Thomson cross section $\sigma_T = 6.65 \times 10^{-29}\,\text{m}^2$. For a central source of radiation expanding spherically, the local intensity I is related to the overall luminosity L by $I = L/4\pi r^2$, where r is the distance to the central source, which is the same r as that used in the calculation of the gravitational force, provided the radiation is generated by accretion onto the central object. Thus the radiation force is given by $\sigma_T L/(4c\pi r^2)$.

The cross section of the electrons to radiation is much greater than that of the protons, so in this case, the protons can be neglected. In most situations the coulomb interactions between the protons and the electrons are strong enough that the plasma remains neutral, and the outward force on the electrons is balanced by the inward force on the protons, and $GMm_p = L\sigma_T/4\pi c$. These two forces balance when the luminosity is equal to the Eddington luminosity L_{Edd}:

$$L_{Edd} = 4\pi \, GMm_p c / \sigma_T.$$

At this luminosity, the outward force of the radiation matches the inward gravitational force and prevents any additional infall. Interestingly, the Eddington limit does not depend on the radius or density of the accreting object but only on its mass. Thus the Eddington limit can be expressed as a function of mass as follows:

$$L_{Edd} \approx 1.2 \times 10^{38} (M/M_\odot) \text{ ergs}^{-1},$$

where $M_\odot$ is the mass of the Sun, and the composition of the gas is assumed to be pure hydrogen. This limit applies not just to energy generated by accretion—a star that attempts to radiate at greater than its Eddington limit will blow itself apart, so this relation provides an upper limit on the brightness of stars generally.

There are some limitations to the derivation shown. A plasma of fully ionized hydrogen was assumed so there are equal numbers of protons and electrons. If the composition is different, then the mean atomic weight of the ions per electron can be different, leading to a different numerical result. If the gas is not fully ionized, the cross section to radiation can be different as well. For example, if a load of bricks were dumped on a black hole, the cross section of each brick would be its geometric size, and the radiation pressure it absorbed would have to balance the inward gravitational force on the brick—although by the time the bricks got close to the black hole, they would likely be heated up and turned into an ionized plasma.

Beyond these considerations of detail, there is a more profound requirement on the Eddington limit, namely, that both the accretion and the radiation are assumed to be spherically symmetric. If the infall is on only part of the object, and radiation escapes in a different direction, the basic assumptions behind the derivation of the Eddington limit do not apply. As we will see, accretion flows onto black holes are not thought to be spherically symmetric— the infall is much more frequently in the form of a flattened disk. But even when the formal requirements for the derivation of the Eddington limit are not met, the Eddington luminosity and the associated limit on mass accretion rate provide useful reference values that have significant physical consequences for accretion flows. Indeed, the vast majority of accreting black holes for which the mass of the black hole is known are radiating at less than L_{Edd} (some possible exceptions are discussed in chapter 7).

2.2 Standard Accretion Disks

In most cases, gas accretion is not spherically symmetric. Infalling gas, like anything else, typically has some rotational component and thus has some angular momentum (usually denoted by J). As the mass falls in, the distance D to the accretor decreases, and the angular velocity V_{ang} must increase so that the angular momentum $J \propto D \times V$ is conserved. The increasing angular velocity generates a centrifugal force that acts against the inward gravitational force. But the gas rotates within a plane, so this outward-directed force operates only in the plane of the rotation. Consequently, the infalling gas naturally assumes the form of a disk oriented perpendicular to the angular momentum vector **J**.

In principle, any object will establish a stable orbit around a black hole, or any other central gravitational mass, with the orbital period and eccentricity determined by the masses of the two objects, and the total energy and angular momentum of the orbit. In Newtonian mechanics, such orbits are stable, and in the absence of other forces like tides or additional gravitating objects, the orbital parameters will not change in time—indeed, the planets in our own solar system have remained in stable orbits for billions of years. Orbits around a black hole can be just as stable as they are around less exotic objects. Whether an orbit is "Newtonian" in its stability is not related to the nature of the orbiting objects themselves but rather to the size of the orbit. If the *orbit* is large compared with the Schwarzschild radius of the two orbiting objects, the orbit will be Newtonian. If the Sun were suddenly replaced by a black hole of one solar mass, the orbit of the Earth would not change at all. Thus the popular impression that a black hole necessarily sucks in everything around it can be very misleading.

One can imagine a single atom establishing a Newtonian orbit around a black hole (or any other object). If that atom is the only orbiting object, it will remain in such an orbit indefinitely. However, if there is a large quantity of gas, each atom of which interacts with the others, the situation becomes quite different. Whatever angular momentum the gas contains is likely to make the gas orbits align in the same plane. If the orbits of the gas are noncircular, then some gas will be moving inward and some, outward. The gas streams will run into each other and exchange energy and angular momentum, so that orbits will become smoothly circular. Thus orbiting gas will naturally configure itself into a disk in which each atom follows a circular orbit around the central body. Such disks are called *accretion disks*.

In an accretion disk, the orbital velocity at distances far from the black hole is determined by standard Newtonian physics, which requires that the centrifugal acceleration V^2/D, plus any other outward forces, be balanced by the inward gravitational pull from the accreting object (GM/D^2). For infall rates well below the Eddington limit, this requirement leads to the Keplerian formula for the orbital velocity $V = \sqrt{GM/D}$, and the familiar result that orbits closer to the central object must be faster. Consequently in a gas disk, this results in friction between gas at any particular radius and the gas immediately closer or farther out, as the particles in the gas slide past each other. This friction generates heat—in formal terms, the viscosity of the gas transfers energy from the kinetic energy of the gas flow to heat energy in the gas. As the viscosity removes energy from the bulk motion of the gas into the random motion of the gas particles it also transports angular momentum outward in the disk. As the bulk kinetic energy decreases, and angular momentum is transported away, the orbiting gas must move into orbits closer and closer to the central object. Thus there is a steady flow of material through the disk toward the central accreting object.

Under some circumstances, the structure of accretion disks can be simply described (see section 2.7). Assuming that the disk is geometrically thin and optically thick, and that the energy dissipated in the disk is radiated thermally, the energy radiated at some radius R in the disk must equal the energy generated by the inward motion of the material through the disk at that radius. Thus

$$\sigma T^4 R dr = \dot{m} dr \, GM/R^2,$$

where σ is the Stefan-Boltzmann contant, and $\dot{m} = dm/dr$ is the rate of mass flow inward through the disk. Thus

$$T \propto R^{-3/4}.$$

The total emitted spectrum of the disk is then the integral of the blackbody emission from the outer to the inner edge of the disk (note that the inner and outer boundaries of the disk generate special conditions). The resulting spectrum has a cutoff similar to a single-temperature blackbody at both the high-energy and the low-energy end but a wider peak (due to the range of temperatures available).

The total luminosity emitted by an accretion disk depends on the rate at which mass is accreting, $\dot{m}$. When the luminosity approaches the Eddington limit, radiation pressure starts to contribute significantly to holding the gas up against gravity, and the thin disk–like geometry can no longer be maintained. Under these circumstance, the accretion flow takes on other geometries. Interestingly, the temperature associated with the peak emission of the disk near the Eddington limit scales with the mass of the accreting object as $T \propto M^{-1/4}$, so more massive black holes tend to produce cooler disks. Of course, the Eddington limit does not strictly apply to nonspherical structures like disks. Nevertheless, the assumption of a thin-disk geometry modified by radiation pressure at luminosities close to the Eddington limit provides reasonable explanations for a surprisingly wide range of observational data.

This approach to accretion disk structure takes the mass accretion rate as a free parameter imposed by the physical situation that generates the gas flow. The assumption is that the disk will adjust itself so that the mass flow is constant throughout the disk—if $\dot{m}$ is lower at some radius of the disk than it is farther out, then mass will accumulate at that radius. The higher density will then generate more friction, more energy loss, and a higher mass flow rate. This process leads, in principle, to a situation in which the local value of $\dot{m}$ is the same at all radii in the disk. To understand the density distribution in the disk, one must then make some assumption about *how* energy is transformed from kinetic energy associated with the rotation flow into heat and radiation, and in particular how this energy transformation processes depend on the density, temperature, and vertical structure of the disk at different distances from the black hole. One assumption, discussed in detail in section 2.7, is that the viscosity that creates the gas friction can be described as a constant α multiplied by the sound speed and scale height of the disk. This assumption leads to a complete physical description of the structure of the disk. Such disks are referred to as α-*disks*.

But this approach does not identify the physical process responsible for the viscosity—it simply assumes a particular functional form for the viscosity. The absence of any notion of the actual physical effects that are taking place makes it all the more remarkable that α-disks really do seem to describe the overall features of real physical objects. However, as observations have improved, many details have emerged that require models that go beyond the α-disk approach and thus necessitate a clear understanding of how the viscosity is generated.

It turns out that the electromagnetic forces between individual ions in the disk gas are much too small to generate the observed flows of mass and angular momentum. Turbulence in the fluid acts to enhance viscosity, but for a long time it was not clear exactly what would drive the turbulence. Recently, it has become generally accepted that a magneto-rotational instability is most likely responsible for the viscosity in accretion disks. In this approach, it is assumed that a magnetic field permeates the disk. Magnetic tension then connects gas orbiting at a given radius with gas slightly farther in (or out). This tension can be thought of as a spring that connects gas particles with other particles at different distances from the black hole. Since the inner gas is orbiting faster than the outer gas, the inner gas pulls the outer gas along, thus transferring angular momentum from the inside of the disk outward, while the outer gas slows down the inner gas, extracting energy from the orbital motion. Detailed simulations of this mechanism have been carried out which demonstrate that it results in disks similar to α-disks in which $0.01 < \alpha < 0.1$, comparable to the values required by observations.

2.3 Radiatively Inefficient Accretion Flows

Not all accretion flows can be described as α-disks. If the various assumptions that give rise to α-disks are violated, quite different accretion flow geometries can be established. Considerable recent research has demonstrated the possibility of a wide variety of accretion flows. While all these models have not been confirmed in observed accreting black holes, some observations that cannot be explained by simple accretion disks have been associated with other kinds of accretion flows.

One important class of accretion flows is the so-called ADAFs, or advection dominated accretion flows. In an ADAF, the assumption that the gravitational energy dissipated by the infalling matter is largely radiated where it is generated is no longer valid. Instead, the gas carries (advects) the energy inward, resulting in a flow that moves inward relatively quickly and is not confined to a plane. At the inner boundary of the accretion flow the gas is likely to have considerable internal energy that was been picked up in the fall inward but has not yet been radiated away. In most situations, the gas flow terminates at a surface, and the energy is transformed into radiation in a boundary layer associated with that surface. However, accretion onto a black hole does not encounter a surface but rather an event horizon. The internal energy of the accretion flow is not liberated at the event horizon but can advect through the event horizon along with the mass that carries it. Thus one of the observational characteristics of accretion onto a black hole is the absence of boundary layer radiation.

One situation in which ADAFs are likely to occur involves very low values of $\dot{m}$, which result in low-density gas flows. In an ionized gas of sufficiently low density, the coupling between positively charged ions and negatively charged electrons through coulomb interactions breaks down, and the ions and electrons can have different temperatures. Since the ions are much more massive than electrons, most of the kinetic energy is carried by the ions. However, the electrons radiate away their energy more efficiently. If the electrons radiate energy faster than it can be replaced by interactions with the ions, then the electrons will be much cooler than the ions, and the ions will advect their energy inward as they accrete onto the central object. Such an accretion flow is necessarily optically thin, and so the observed radiation field is not thermal, as is the case for standard accretion disks, but rather some combination of nonthermal processes.

2.4 Accretion Instabilities

We have assumed so far that the accretion rate and other characteristics of the accretion disk do not change in time. However, accretion flows are observed to vary, sometimes quite drastically. A typical light curve showing intensity

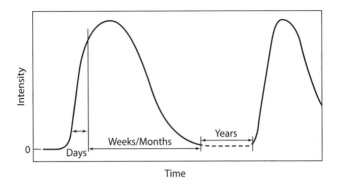

Figure 2.1. X-ray binaries often display a *fast rise and exponential decay* (FRED) light curve like the one shown. In some cases sources can go from invisible to among the brightest sources in the sky within a few days.

versus time for an unstable accreting stellar mass black hole is shown in figure 2.1. There are in general two causes of variability in disks and other accretion flows. The first results from changes in the overall mass accretion rate $\dot{m}$. Such a change implies that the physical situation in which the accretor is embedded has changed. Since the energetics of accretion depend on how much mass is being accreted, such changes result in changes in the physical conditions and observed properties of the accretion flow.

However, accretion flows can be unstable even if the overall accretion rate is constant, as a result of instabilities in the gas flow. In the same way that water can drip from a faucet rather than flowing continuously, the gas flow *through* the disk can be variable or episodic, even if the gas flow onto the disk is constant and continuous. One situation that has been explored in detail involves changes in the ionization of the gas. The free electrons and ions in ionized gas generate a significantly higher viscosity than neutral gas, and thus the local $\dot{m}$ will be greater when the gas is ionized than when it is not. If the mass entering some region of the disk is less than the disk can process when the gas is ionized, but more than moves through the disk when the gas is neutral, then a *limit cycle* can be set up. In such a limit cycle, the gas density increases when the gas is neutral, since the mass accretion rate is greater than the rate at which gas is moving through that location of the disk. As the gas accumulates, it gets denser and hotter, until ionization occurs, which greatly increases the viscosity of the disk. Then, the accumulated gas "flushes" rapidly down through that region of the disk, leading in turn to a rapid drop in density and temperature, and a recombination of the gas. The cycle then begins anew. In the α-disk formalism, such disk instabilities can be modeled as abrupt changes in the assumed value for α.

One consequence of a change in either the overall mass accretion rate or the mass transfer rate through different parts of the accretion flow is that the geometry of the flow can change dramatically. Objects have been observed to change from α-disks to ADAFs and back, and some situations seem to require different kinds of flows at different distances from the accreting object. Such changes in the accretion process are accompanied by significant changes in the observed spectral properties of the object, so changes in the mass accretion rate do not change merely the overall luminosity of the source but also the spectrum and other detailed characteristics of the emerging radiation. Such changes in the observed characteristics of accreting objects

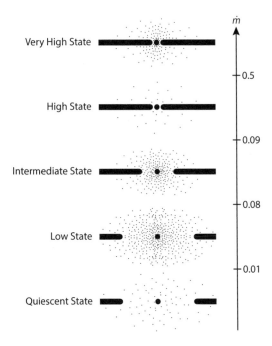

Figure 2.2. Cartoon representation of accretion states in an X-ray binary, where $\dot{m}$ is given in units of the Eddington rate. Note the increased dominance of a disk at higher accretion rates. It is now believed that accretion rate alone is not the only determinant of the accretion flow. From A. Esin, J. McClintock, and R. Narayan, 1997, *Astrophys J.*, 489: 865.

are referred to as *state transitions*. One hypothezised set of state transitions as $\dot{M}$ changes is shown in figure 2.2.

2.5 Radiation Emission Mechanisms

Accretion flows onto a black hole cannot be mapped directly. The angular scale projected on the sky by the event horizon and its environs is generally much too small to be resolved by current instrumentation. Therefore, what we observe is a single point source of emission in which the radiation from the whole accretion flow is combined. Thus we have to infer the geometry of the accretion flow from the spectrum and variability of the total observed emission. To do this we must understand how the energy generated by the infall of material is transformed into radiation (the *emission mechanisms*) and how that radiation propagates from where it is produced to where it is observed (*radiative transfer*).

The simplest emission mechanism occurs when an object is optically thick, that is to say, when the photons are repeatedly scattered by the material so that they eventually emerge only from the surface of the object, and their properties are determined solely by the temperature of the emitting surface. Such radiation is referred to as *blackbody radiation*. As noted earlier, the emitted spectrum of an optically thick accretion disk can be simply determined by integrating the blackbody spectrum from the hot inner edge of the accretion disk to the cooler outer edge. Such a spectrum has an exponential cutoff at wavelengths shorter than those associated with the temperature at the inner edge of the disk, and a power-law cutoff at wavelengths longer than those associated with the temperature at the outer edge of the disk. In between, the spectrum is flatter than that of a single-temperature blackbody. Nevertheless, one can compute a wavelength region where most of the emission is expected to be emitted. This wavelength varies with the mass of the accreting object as $\lambda_{\max} \propto M^{1/4}$. Since the temperature of a blackbody is inversely proportional to the wavelength where the maximum radiation is emitted, this means that the inferred temperature for an accretion disk varies as $T \propto M^{-1/4}$. For accreting masses similar to those of stars, the radiation peaks in the soft (low-energy) X-rays, as is observed from the strong galactic X-ray sources. This agreement of theory and observation helped establish the nature of the galactic sources as accreting black holes and validated the general disk accretion hypothesis. For much more massive black holes, the radiation should peak in the ultraviolet, as is also observed.

In other kinds of accretion flows, the accreting material is often optically thin, and in this case the radiation emission mechanisms can be more varied and more complex.[1] Generally speaking, the emission arises from acceleration of electrons by various mechanisms; acceleration of electric charges generates electromagnetic radiation, and the electrons, being much less massive than protons or other charged particles, are particularly prone to acceleration. The acceleration can be generated by interactions with other charged particles, by magnetic fields, or by interactions with photons.

Bremsstrahlung. Bremsstrahlung radiation is emitted when an electron is deflected by the electromagnetic charge of some other particle. This process is common in ionized plasmas, where there are many free electrons and also large numbers of positively charged ions. When the electrons are unbound both before and after they interact with the ions, the radiation is sometimes referred to as *free-free emission.*

---

[1] See the classic textbook by G. Rybicki and A. Lightman, *Radiative Processes in Astrophysics* (New York: Wiley, 1979) for a full discussion.

The wavelengths and intensity of the resulting radiation depend on the distribution and velocity of the electrons and ions, and thus on the temperature of the plasma. The specific case in which the electrons all have the same temperature is called *thermal bremsstrahlung*.

Synchrotron emission. In the presence of magnetic fields, the electrons circle the magnetic field lines. Nonrelativistic electrons generate *cyclotron emission*, at a frequency determined by the energy of the electrons and the strength of the magnetic field. Relativistic electrons complicate the situation somewhat, since the resulting radiation is beamed forward, rather than being emitted isotropically. The resulting radiation is referred to as *synchrotron emission*. In general, an emitting plasma contains electrons with a distribution of energies interacting with magnetic fields of a variety of strengths and directions. The resulting spectrum can often be described by an intensity distribution that varies as a power law with frequency. Radio emission from accreting black holes is often due to synchrotron emission.

Compton scattering. Compton scattering occurs when electrons and photons interact. Energy can be exchanged between the photons and electrons, resulting in dramatic changes to the observed spectrum of photons. In particular, a seed population of photons interacting with a distribution of relativistic electrons can be scattered to high energies. One common situation is referred to as *synchrotron self-Compton* or SSC emission. In this configuration, an initial seed spectrum of low-energy radiation is generated by the synchrotron process, and then the same relativistic electrons that initially generated the photons also generate much higher energy radiation through the Compton process.

Line emission. Emission confined to specific wavelengths (emission lines) can be created and removed by atomic and nuclear processes. When an electron falls from a high energy level to a lower energy level, a photon of a specific energy, and thus a specific wavelength, is released. If some process returns the atoms to the higher energy level, a steady stream of photons at a specific wavelength can be produced. Several processes can repopulate the higher energy levels, including collisions or energy input from a radiation field.

As a plasma is heated the gas becomes increasingly ionized, which changes the atomic transitions responsible for line emission, so the high-temperature plasmas near black holes have spectral lines that are quite different from those commonly observed in stars. A fully ionized plasma does not generate atomic line emission at all. The highest-energy atomic line generally observed comes from 25-times-ionized iron (that is, an iron nucleus surrounded by a single electron), which generates a hydrogen-like spectrum, but at $26^2 = 676$ times higher energy, given that the energy levels of single-electron atoms scale as the square of the nuclear charge. The equivalent of the Lyman-α line in hydrogen (produced by electrons falling from the second to the lowest energy level) occurs at 6.4 keV in 25-times-ionized iron. X-ray telescopes can easily observe this energy, and since iron is one of the more commonly occurring heavy elements in astrophysical plasmas, this line is often observed and studied. It is the highest energy at which atomic transition is generally observed. However, other processes can generate lines at higher energies. Nuclear processes and radioactivity can give rise to line emission, and annihilation of electrons and positrons generates radiation at 511 keV.

Line emission from a disk (or from an optically thin atmosphere above a disk) has a particular shape due to the Doppler shift generated by the motions in the disk. Such a line will be broadened by the range of Doppler shifts and will appear to have two peaks, generated by the parts of the disk that are approaching and receding from the observer. Such double-peaked emission lines are commonly observed and are generally considered to be clear evidence for the presence of a rapidly rotating disk.

2.6 Radiative Transfer

Once the radiation has been generated, it must travel to the observer. The observed spectrum can be changed significantly by interactions between the radiation and material it passes through. The relevant physical processes are referred to as *radiative transfer*. Changes in the spectrum due to radiative transfer will naturally occur as the radiation field emerges from the source—all the processes already described are examples of how interactions with an ambient medium can affect an emerging radiation field. But the radiation field can also be absorbed by material far from the original source, along the line of sight to the observer. Since the amount of absorption depends on the wavelength of the radiation, such interstellar absorption can dramatically change the observed spectrum. Crucial sources of absorption include the following.

1) Dust. Interstellar dust particles can absorb radiation in ultraviolet, optical, and infrared wavelengths. Dust generally absorbs shorter wavelengths (bluer light) more than longer wavelengths. In the optical, such absorption makes the spectrum look redder, so dust absorption is sometimes referred to as *interstellar reddening*. There are standard models of dust that predict the amount of absorption as a function of wavelength and column density of the dust, and these models work well under most circumstances.

2) Hydrogen ionization. Hydrogen is by far the most common element in the Universe, and the interstellar medium has a significant density of hydrogen. Most of this hydrogen is cold, with its single electron in the lowest energy state. This interstellar hydrogen absorbs photons whose energy is sufficient to ionize ground-state hydrogen. Thus ultraviolet radiation below 912 Å (>13.6 eV in photon energy) generally does not propagate successfully through the galaxy. At much higher energies the cross section of hydrogen becomes smaller, so that for soft X-rays at >1 keV the effect of hydrogen absorption is greatly reduced. But the extreme

ultraviolet, where hydrogen ionization has a large effect, is a very difficult wavelength regime to observe. The amount of interstellar hydrogen scales roughly with the amount of interstellar dust, so optical reddening and hydrogen absorption of ultraviolet and X-ray photons are often correlated.

3) At the highest energy, photons with energies ≥ 10 TeV interact with the low-energy photons of the pervasive microwave background to create electron-positron pairs. This process severely limits the propagation of very high energy photons across cosmological distances.

All these physical processes are observed in accreting black holes, often simultaneously. Thus understanding the radiation spectrum generated by accreting black holes often requires a superposition of thermal emission, nonthermal emission (often approximated as a power law), and line emission, as well as a consideration of the effects of interstellar absorption. Often, different components of the spectrum, and different wavelength regimes, are generated in different parts of the accretion flow. We observe thermal optical emission from a standard accretion disk, radio emission generated by synchrotron emission, and high-energy emission from comptonization all in the same object. It is not always straightforward to correctly decompose an observed spectrum into components associated with specific emission mechanisms. It can also be difficult to associate specific accretion flow geometries with emission mechanisms and the resulting observed spectrum. Ideally, one would create fully self-consistent models of observed spectra, taking into account the energy generated by the accretion, all the processes that transform this energy into observable radiation, and all relevant radiative transfer effects. In practice, this is difficult to achieve, so often, assumptions are made about the accretion flow or the radiation mechanisms that are not fully justified by the observations.

2.7 The α-Disk

The α-*disk* or *thin-disk* model for accretion disks was first worked out by Shakura and Sunyaev.[2] In essence, a set of assumptions about the nature of the accretion process is made that leads to a set of eight algebraic equations in eight unknowns. These are the numbered equations that follow. These equations can then be solved as a function of the distance from the accreting object R for any particular value of the accreting mass M and the mass accretion rate $\dot{m}$. The assumptions made are by no means always applicable, so accretion flows can be dramatically different in nature from α-disks. Nevertheless, the α-disk formulation is useful for many applications.

The first key assumption is that all elements of the disk travel in Keplerian circular orbits around the accreting objects. In this case one can write the equation of circular

[2]N. Shakura and R. Sunyaev, 1973, *Astronomy & Astrophysics* 24:337. Here we use the notation of J. Frank, A. King, and D. Raine in chapter 5 of *Accretion Power in Astrophysics* (Cambridge: Cambridge University Press, 3rd ed., 2002), where further comments and details of the derivations can be found.

motion $V_{\text{circ}}(R) = R\Omega(R) = (GM/R)^{1/2}$, where R is
the distance of a particular gas element from the accreting
object, and Ω is the angular velocity of the gas. Note
that this equation also assumes that the gravity from the
accreting object dominates any gravitational effect of the
companion star or the disk itself.

However, if this assumption were strictly true, the gas
would simply continue in stable orbits, and no accretion
would take place. The assumption neglects the viscosity
of the gas. Viscosity provides a way to transfer angular
momentum and energy within the disk, allowing the gas
to move from one orbit to another, and thus drives the
accretion process. One way to think about the effects of
viscosity is to consider that gas in a circular orbit at distance
R from a central gravitating object has a greater angular
velocity than gas orbiting at a greater distance. Viscosity
creates tension between gas at different distances from the
accreting object, which results in the outer gas being pulled
forward by the more rapidly orbiting inner gas, while the
inner gas is slowed down by the outer gas. Thus angular
momentum moves from the inner gas outward, while the
inner gas tends to slow down and thus spiral inward. As
the gas moves toward the center the total energy of the gas
decreases, and the lost energy increases the internal energy
of the disk itself and is ultimately radiated away. It is this
radiation that is ultimately observed.

Given the presence of viscosity, the circular orbit equation is an approximation, and its use assumes that the
inward drift of the gas in the radial direction v_R is small
compared with the orbital speeds, that is, $v_R \ll V_{\text{circ}}$.
The mass transfer rate $\dot{M}$ for the gas can then be written

$\dot{M} = 2\pi R \Sigma v_R$, where $\Sigma(R)$ is the surface density of the gas, and thus the total mass at any given annulus is $2\pi R \Sigma d R$. As the mass moves in toward the center of the system, mass, angular momentum, and energy must be conserved. Each of these conservation laws leads to an equation that relates the mass flow through an annulus at distance R from the accreting object to v_R and the conditions in the disk at that point.

In writing these three equations, we make two additional crucial assumptions. First, we assume that the disk is in a steady state, and thus its structure does not change with time. Second, we assume that the disk is thin, that is, that its vertical scale height $H(R)$ is everywhere small compared with R. More specifically, H relates the midplane density $\rho(R)$ at any point in the disk to the surface density Σ by

$$(1) \qquad \rho = \Sigma/H.$$

Furthermore, for any quantity X that varies with vertical distance above the disk z, $\delta X/\delta z \approx X/H$. A "thin" disk implies $H \approx z << R$. For example, the pressure gradient $\delta P/\delta z$ can be written P/H. In the vertical direction, the pressure gradient must be balanced by the z component of the gravitational force from the central object, so that

$$\frac{\delta P}{\delta z}(1/\rho) = \frac{-GMz}{R^3}.$$

If we use the equation of state

$$(2) \qquad P = \rho c_s^2,$$

where c_s is the sound speed, we find that

$$(3) \qquad H = c_s \left(\frac{R^3}{GM} \right)^{1/2}.$$

The last equation, the equation of state, and the definition of Σ give us three of our eight equations of disk structure.

Another relevant equation can be found by setting the pressure equal to the sum of the gas pressure $\rho k T_c / \mu m_p$, where k is the Boltzmann constant, T_c is the central temperature of the disk, μ is the atomic weight of the gas, and m_p is the mass of the proton; and the radiation pressure, given by $(4\sigma/3c) T_c^4$, where σ is the Stefan-Boltzmann constant:

$$(4) \qquad P = \frac{\rho K T_c}{\mu m_p} + \frac{4\sigma}{3c} T_c^4.$$

Conservation of mass can be simply expressed as

$$\frac{\delta(R\Sigma v_R)}{\delta R} = 0.$$

$R\Sigma$ is proportional to the amount of mass in an annulus at R, and v_R is the rate at which the mass is flowing toward the interior, so this equation simply states that there can be no net flow of mass from one part of the disk to another—mass being transported inward must be balanced by mass flowing in from the outside. If this were not true, the total amount of mass at some point in the disk would be changing, thus violating the assumption of a steady-state disk.

The equations of angular momentum conservation and energy conservation are more complex, because the viscosity provides a local source for these quantities. The angular momentum of the gas at R can be written as $2\pi R \Sigma R^2 \Omega$, so conservation of angular momentum can be written

$$\frac{\delta(2\pi R \Sigma R^2 \Omega v_R)}{\delta R} = \frac{\delta Q}{\delta R},$$

where $Q(R)$ is the torque exerted by the viscosity. Note that at an annulus at R the *net* torque is the difference between the torque exerted by gas at $R + \delta$ on that at R, and that of the gas at R on that at $R - \delta R$. This is why the right-hand side of the equation takes the form of a derivative of the torque. Here we ignore the constant of integration, which relates to the angular momentum transfer between the inner edge of the accretion disk and the accreting object, and simply set Q equal to the quantity in parentheses.

If the torque Q is caused by viscosity across neighboring gas annuli, that is, if it is a purely local effect, it must be proportional to the strength of the viscosity v, and the change in the angular velocity across neighboring annuli $d\Omega/dR$, and can be written $Q(R) = Rv\Sigma R^2(d\Omega/dR)$. If we now assume Keplerian rotation and use the definition of $\dot{M}$, we find that angular momentum conservation in the disk can be expressed as

$$(5) \qquad v\Sigma = \dot{M}/(3\pi).$$

This is a reasonable result, in that it says that the mass transfer rate increases with the density of matter in the disk (Σ) and with the strength of the viscosity, which is the physical process that leads to the mass transfer in the first place.

Similarly, the energy deposited in the annulus at R can be written as a luminosity $L = Q\Omega' = 3GM\dot{M}/(2R^2)$, where we have used the expressions for Q, $\dot{M}$, and Ω given previously. This energy then must be radiated away from the disk. The radiant blackbody flux emerging from a thin disk can be written $F = 4\sigma T_c^4/3\tau$, where the thin-disk approximation is used to relate the optical thickness of the disk τ and the central temperature T_c to the temperature appropriate for the blackbody radiation, which is the temperature at $\tau = 1$. The flux emerges from both sides of the disk, and thus over an annulus is equal to $4\pi RF$, with F as given earlier. This expression can then be set equal to the luminosity generated by the viscosity to yield

$$(6) \qquad \frac{4\sigma T_c^4}{3\tau} = \frac{3GM\dot{M}}{8\pi R^3}.$$

Note that this assumes that the energy is transported by radiation and that the disk is optically thick—otherwise the assumption of blackbody radiation does not hold. The optical depth can be determined from the state of the gas and is generally taken to be well represented by Kramer's opacity, in which case

$$(7) \qquad \tau = \rho H \kappa_R = C \Sigma \rho T_c^{-7/2},$$

where C is a constant determined by the Gaunt factors associated with quantum mechanical effects and the composition and ionization state of the gas.

We now have a set of seven equations in the eight unknowns ρ, Σ, H, c_s, P, T_c, τ, and ν, all of which are functions of the mass of the accreting object M, the mass accretion rate $\dot{M}$, and the annular distance from the accreting source R. To close the system, we need an expression for the viscosity in terms of the other parameters. The final assumption then is then that

$$(8) \qquad \nu = \alpha c_s H,$$

where α is a constant, generally taken to be less than 1. This assumption, first used by Shakura and Sunyaev, produces a set of eight algebraic equations in eight unknowns, which can then be solved to determine the density, temperature, and emitted spectrum for every value of R. Integration of the emission over $2\pi\, R\, dR$ then produces a spectrum for the disk as a whole. Thus for any value of M and $\dot{M}$, the spectrum produced by the accretion disk can be calculated without solving any differential equations at all.

This calculation contains a large number of assumptions, which we collect here for reference:

- The gas orbits in near-circular orbits; that is, $v_R << V_{\text{circ}}$.
- The disk is in a steady state with a mass accretion rate $\dot{M}$ that is constant in time and across the disk.
- The disk is geometrically thin and optically thick.
- The details of the inner boundary, where mass accumulates on the accreting object, are unimportant (note that this assumption can be

relaxed and the equations rederived—see Frank, King, and Raine).

- Viscosity is local and can be approximated by $\nu = \alpha c_s H$.

Remarkably, many (but by no means all) accreting objects have spectra that resemble those predicted by this simple theory. This fact, along with the ease with which the emission and physical properties of these α-disks can be calculated, has led to the extensive use of α-disks as the first, most obvious approximation of disk structure for a large range of research on accretion disks over the past 40 years.

3

OUTFLOWS AND JETS

One somewhat unexpected feature of accretion flows is the presence of outflows, or jets. There is strong observational evidence that some fraction of the infalling material reverses course near the accreting object and is shot out perpendicularly to the accretion disk. In some cases outflow velocities in the jet can be close to the speed of light, and the jets can carry energy over very large distances. In particular, narrow collimated beams of emission are observed emerging from the central-most regions of galaxies and continuing across the whole of the galaxy, depositing their energy hundreds of kiloparsecs away from their origin (see figure 3.1). Blobs of material are observed to move across the sky at relativistic speeds. These phenomena are sometimes described as jets "emerging" from a black hole. This parlance is misleading—the jets do not, and indeed could not, emerge from inside the event horizon. Rather, some mechanism redirects the energy generated by the accretion process into a fraction of the infalling material and provides enough bulk kinetic energy for the material to escape the accretion process before the material enters the event horizon. This energy is eventually deposited into the regions surrounding the black hole, often at large

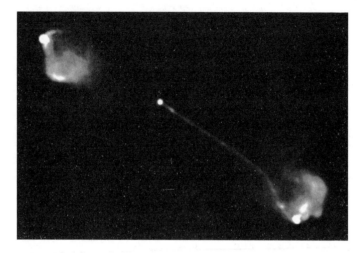

Figure 3.1. Radio image of the quasar 3C175 obtained at the Very Large Array (VLA) at a wavelength of 6 cm. The AGN is at the central bright spot, and the radio lobes are hundreds of kiloparsecs away. In this case a one-dimensional jet connecting the AGN and the lobe is clearly seen. Image copyright to the National Radio Astronomy Observatory (NRAO) 1996.

distances from the black hole itself. In this way the black hole can influence its surroundings in ways other than by gravitational attraction.

Various kinds of accretion processes unrelated to black holes also generate jets. For example, accretion in a variety of binary stars is known to result in jets. Young stellar objects that are still accreting gas in an accretion disk from the protostellar cloud that formed them also generate jets. Thus it appears that disk accretion often leads to redirection of material in the vertical direction. This material tends to emerge at velocities comparable to the escape velocity of the accreting object. This suggests that the physical mechanisms involved in jet production occur close to the surface of the accreting object and are associated with the accretion process in general rather than being specific to the nature of the particular accretor. Consequently, the existence of jets emerging from accreting objects at close to the speed of light implies that the accreting object itself must have an escape velocity close to c and thus is likely to be a black hole.

3.1 Superluminal Motion

Jets associated with black hole accretion often appear to move across the sky with a velocity v_{obs} that appears to be greater than the speed of light. That is, the angular velocity of jet material across the sky $\dot{\omega}$ multiplied by the distance to the object D, can yield a value $D\dot{\omega} = v_{obs} > c$. Such jets are known as *superluminal jets*. The apparent superluminal motion is an illusion—the true space velocity v of the material in the jet is necessarily less than c.

The apparent superluminal motion is a combination of two effects. As the jet approaches observer the light travel time to the observer decreases, so the observed time for the jet to travel a certain distance from the source is decreased by the change in the light travel time. This decrease is coupled with time dilation due to special relativity (see figure 3.2). These effects combine, such that

$$\beta_{obs} = \frac{\beta \sin \theta}{1 - \beta \cos \theta},$$

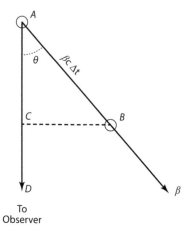

Figure 3.2. Geometry of a superluminal jet. If an emitting region moves from A to B in time Δt at a speed approaching the speed of light (that is $\beta = v/c \sim 1$) and the angle θ is small, its transverse velocity will appear to be greater than the speed of light.

where $\beta_{obs} = v_{obs}/c$, $\beta = v/c$, and θ is the angle between the direction of jet propagation and the line of sight. It can easily be seen that β_{obs} can sometimes be greater than unity, that is, that the observed motion of a blob emitted at relativistic speeds across the sky can appear to be greater than the speed of light, even if $\beta < 1$. The maximum value of β_{obs} is obtained for $\cos\theta = \beta$, in which case $\beta_{obs} = \beta/\sqrt{1 - \beta^2}$. There are many examples of such apparently superluminal motion associated with both galactic and extragalactic black holes.

Special relativity can also create dramatic changes in the observed intensity, variability, and energy of the radiation from the jet. The critical quantity is $\delta = (1 - \beta\cos\theta)^{-1}$, which ranges from unity for nonrelativistic flows to infinity for a flow directly toward the observer at the speed of light. In terms of δ, the observed timescales in the source diminish as $t_{obs} = t/\delta$, and so the observed frequencies increase as $\nu_{obs} = \nu\delta$. Thus the energy of the photons in the observed frame is greater than that of those in the emitted frame. In addition to these effects, an isotropic radiation field is beamed, so that half the photons are concentrated into a cone with opening angle θ in the forward direction, where θ is given by $\sin\theta = 1/\gamma$, where γ is the Lorentz factor $\gamma = 1/\sqrt{1 - \beta^2}$.

The consequence of these effects is that the overall luminosity of the jet goes as $L_{obs} = L\delta^4$, and is thus greatly enhanced, as well as bluer and more rapidly varying, in a jet directed toward the observer compared with the luminosity of an otherwise equivalent stationary source.[1] Thus sources in which the jet is beamed toward Earth provide particularly good observations of jets, since the power of the jet is enhanced in comparison with that of stationary sources of emission. The class of extragalactic objects for which the jet emission dominates other radiation is known as *blazars*—these sources emit some of the most luminous and highest energy radiation yet observed.

---

[1] Derivations of these results from the Lorentz transformations are an excellent exercise for students and can be found in many textbooks and lecture notes. A useful discussion of the relationship between the textbook special relativity results and observations of relativistic jets is given by G. Ghisellini in "Special Relativity at Action in the Universe" in B. Casciaro, D. Fortunato, M. Francaviglia, and A. Masiello (eds.), *Recent Developments in General Relativity* (Milano: Springer-Verlag, 2000), 5.

3.2 Jet Physics and Magnetohydrodynamics

Some explanations of relativistic jets rely on the relativistic effects of a rotating Kerr black hole to collimate and redirect the flow. Such explanations imply that the spin of the black hole powers the jet, but spin measurements in accreting stellar black holes cast some empirical doubt on this argument (see chapter 8). This class of explanations is also presumably not relevant to the nonrelativistic jets seen in other kinds of systems.

A more widely accepted model that could apply to other accretion-powered jets is that magnetic fields in the inner accretion disk play the critical role. The study of the interplay of fluid flows and magnetic fields is called *magnetohydrodynamics* (MHD), the basic equations of which can be derived from considerations of conservation of mass, momentum, angular momentum, and energy, in a manner similar to that for nonmagnetic fluid flows. These equations are combined with Maxwell's equations to generate a set of differential equations that describe both the motion of the fluid and the topology and evolution of the magnetic field.

These equations can only be solved analytically in very simplified cases. However, insight into the basic physics can be found from considering two nondimensional numbers. The first is the *magnetic Reynolds number* $R_m = VL/\eta$, where V and L are typical velocities and length scales associated with the flow and the magnetic field evolution, and η is the magnetic diffusivity. High values of R_m often prevail in astrophysical settings, in which case the magnetic field is "frozen in" in the fluid. That is, a particular piece of the fluid is attached to a particular magnetic field line. As the gas flows and the field evolves, the field and the flow remain attached to each other, and the magnetic field does not diffuse across the velocity field of the fluid, and vice versa. This situation is sometimes called *ideal MHD*.

Ideal MHD can sometimes result in a very tangled magnetic field as a complicated fluid flow ties the field up in knots, which, in turn, dramatically decreases the length scale L. One way to consider L is as the length across which the magnetic field changes direction or strength by a significant amount. If the field becomes knotted, L is the length across one strand of the knot. As the gas continues to flow, carrying the field lines with it, the knots can become tighter and tighter, L decreases, and R_m decreases along with it. Eventually, R_m becomes sufficiently small that the conditions for ideal MHD no longer apply. Then, magnetic diffusion becomes important, and the magnetic field lines can slip across the fluid. This slippage results in *magnetic reconnection*, in which the field lines reattach themselves into simpler configurations, and complex knotted field lines are resolved into simpler topologies. This simplification is accompanied by a release of the energy associated with the magnetic field.

The other key parameter is the ratio of the gas pressure to the magnetic pressure, generally denoted by $\beta = nkT/(2\mu_0 B^2)$, where nkT is the gas pressure, μ_0 is the magnetic permeability of the vacuum, and B is the magnetic field strength. Note that this β is not to be confused with the relativistic parameter $\beta = v/c$. If $\beta \gg 1$, the gas pressure dominates. In this case, for ideal MHD the motion of the gas is dominated by hydrodynamics, and the magnetic fields are pulled along by the movement of the gas, as described earlier. However, if $\beta \ll 1$, the flow is determined by the magnetic field, and the gas is forced to follow the magnetic field lines, like beads on a wire. In this case, the magnetic field configuration determines where the gas flows. It is thought that the collimation and confinement of the jet to a narrow stream is a result of a situation where the magnetic pressure dominates, and the field lines form a "nozzle" confining the flow.

Thus the basic scenario for jet formation is that in the differentially rotating accretion disk, the magnetic fields are pulled along and tend to wind up, becoming ever more powerful. At some point the magnetic field reconnects, breaking free of the disk, creating very energetic flares that may be the trigger for jet events. The jet is collimated by the wound-up magnetic field, which constrains the outward flow of material to a tight beam. As the jet propagates away from the accreting object, hydrodynamic and magnetohydrodynamic effects generate the observed bends, twists, and knots in the flow. These ideas can in principle account for the formation, collimation, and propagation of jets.

Each piece of this scenario can be studied using simplified approximations that can be explored analytically. But while this approach can verify that these ideas might work in principle, they are not sufficiently detailed to compare directly with the observations—computer simulations are required. Such simulations generally take one of two approaches. The first is a *smooth particle hydrodynamics* (SPH) approach, in which the gas is divided into small point masses, and the motion of each is subject to gravity, gas pressure, and magnetic pressure generated by the other points. Then, the trajectories of all the points are followed, providing a simulation of the gas flow. Algorithms of this type are often referred to as *Lagrangian* approaches. The alternative *Eulerian* approach is to divide the space into cells, each of which is assigned a gas density and temperature, a magnetic field strength and direction, and other thermodynamic variables. Then, the computation determines how the properties of each cell change with time under the influence of the various forces.

Whatever approach is used, detailed calculations of the kind necessary to compare results with the extensive observations of jets are difficult to carry out. Three-dimensional treatments are necessary, since an inward-flowing disk must be redirected into a perpendicular outward-flowing jet. To cover a substantial volume of space requires a large number of SPH points or Eulerian cells, each of which must be updated at each time step. A wide range of scale sizes must be treated—from microscopic magnetic eddies to bulk flows across hundreds of kiloparsecs of space— which also increases the required number of points or cells. Formulating useful approximations of the physics is also difficult, in that all the various regimes are important in different parts of the system. Ideal MHD with high R_m is appropriate in some cases, but diffusive reconnection is also required to transmit energy to the jet. Gas-dominated flows are needed in the disk to wind up the flow, while magnetic-dominated configurations are important to collimate the flow. Thus realistic simulations of jet physics can be very complex.

However, qualitative advances in the simulations have recently become possible, leading to a new era in the study of jet physics. Increasing computer power has made fully three-dimensional flows tractable. At the same time, more sophisticated algorithms, some of which combine the virtues of Eulerian and Lagrangian approaches, have been developed and put into codes that take advantage of parallel computing hardware. For these reasons, computational magnetohydrodyamics has become a very active field, and a more detailed understanding of jet physics may soon become available.

4

STELLAR-MASS BLACK HOLES

The empirical study of black holes began in the 1960s with the discovery of quasars (discussed in the next chapter) and the advent of X-ray astronomy. Since the atmosphere is opaque, celestial X-rays cannot be observed from the ground, so X-ray astronomy did not begin until the space age. Once observing X-rays became possible, it quickly became clear that the X-ray sky is dramatically different from the sky seen in optical light. Most of the optical light in the sky comes from stars, with nuclear fusion as the underlying power source. It is not surprising that this is the kind of radiation that our eyes have evolved to detect, bathed as we are by radiation from a particular nearby star. But other kinds of radiation have revealed whole new categories of objects, powered by processes that are not important in the Sun, particularly accretion. Indeed, the series of dramatic advances in astronomy from the mid-twentieth century to the present have largely been due to the development of instruments that can detect new forms of radiation. In the case of X-ray astronomy, the brightest sources of celestial X-rays turned out to be accretion-powered binary star systems. In many of these systems, there is compelling evidence that the accreting object is a black hole.

4.1 X-Ray Binaries

In the early 1960s, rockets with X-ray detectors began to be launched above the Earth's atmosphere. These instruments were essentially rocket-mounted Geiger counters that could detect X-rays coming from a particular direction—as the instrument rotated, the detector scanned the sky. The initial motivation was to study solar system phenomena. In particular, the earliest X-ray observatories were designed to detect solar X-rays reflecting from the Moon, in hope of determining the nature of the lunar surface in preparation for the manned missions to the Moon. It was already known that the Sun is an X-ray source, although the total X-ray luminosity is much smaller than the total solar luminosity, which is predominantly optical light. The inferred brightness of other much more distant stars was very low, so it was not expected that X-ray sources from outside the solar system would be detectable.

However, it was quickly discovered that there were strong X-ray sources that appeared in the same position in every scan. Such spatial stability would be expected from celestial sources but not from solar system objects, which one would expect to move by significant amounts relative to the Earth, even during the short rocket flights. The observed sources were not coincident with any of the known nearby stars and thus had to be located at significant cosmic distances. The inferred luminosity of the sources was hundreds or thousands of times brighter than the Sun. When coincident optical stars were identified, they proved to be relatively faint. Thus it was clear that a new class of celestial sources must exist whose radiation is predominantly in the form of X-rays, with a total luminosity comparable to or greater than that of ordinary stars.

Attempts to understand this exciting new X-ray universe led to the launch of more X-ray–sensitive probes, culminating in 1970 with the launch of the first orbiting X-ray observatory, the *Uhuru* satellite. In the years since then, a series of increasingly sophisticated X-ray satellites have been launched, culminating in the currently operating Chandra and XMM-Newton missions. The sensitivity and spatial and spectral resolution of X-ray observatories now rival those at optical and other wavelengths, and X-ray astronomy has become a vital part of virtually all areas of astrophysics.

Uhuru observed and cataloged hundreds of X-ray sources across the sky. The Uhuru sky map is shown in figure 4.1, in galactic coordinates, in which the plane of the galaxy is a horizontal band across the middle of the diagram, and the direction toward the center of the galaxy is in the middle of the plot. It is immediately apparent that the majority of the observed sources must be galactic, since they are concentrated in the galactic plane and in particular toward the galactic center.

Simply identifying the general location of these bright X-ray sources led to a determination of some of their key physical characteristics. In particular, the approximate distance to the X-ray sources must be comparable to the distance to the galactic center, where many of them clearly reside. Since the distance toward the galactic center is known to be around 8 kpc, the luminosity of the sources could be approximately determined from the observed flux F and the known distance by the inverse square law

The Fourth Uhuru Catalog

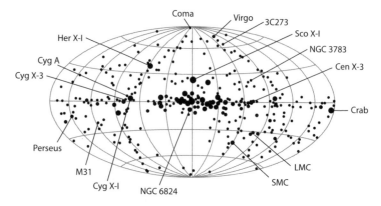

Figure 4.1. Fourth and final catalog of X-ray sources from the *Uhuru* satellite. The sources are displayed in galactic coordinates (with the plane of the galaxy in the equator), and the strength of the source is indicated by the size of the circle. Both galactic and extragalactic sources are clearly detected. Image from NASA, Uhuru Mission.

$F = L/4\pi D^2$. The resulting luminosity proved to be in the range $L \approx 10^{36}$–10^{38} erg s^{-1}, close to the Eddington luminosity associated with stellar masses.

It was also clear that the volume from which the X-rays are emitted must be very small. Two lines of argument led to this conclusion. First, the size of a thermal source of radiation can be determined by its luminosity and temperature, through the Stefan-Boltzmann equation

$$L = \sigma \, 4\pi \, R^2 \, T^4,$$

where L is the luminosity of a thermal source, R is the radius of the emitting region, T is the temperature, and

σ is the Stefan-Boltzmann constant. The temperature of a blackbody is determined by $T\lambda_{max} = 2.9 \times 10^{-3}$, where T is the temperature (in Kelvin) and λ_{max} is the wavelength (in meters) at which the most radiation is emitted. If the radiation is dominated by soft X-rays ($\lambda \approx 10^{-9}$ m), then the blackbody temperature must be in the millions. Plugging this temperature and the observed luminosities into the blackbody formula demonstrates that the emitting region must be smaller than the Earth and much smaller than most stars.

An even stronger argument for a small emitting region came from the very short variability timescales of the emission, which is observed to change significantly on timescales shorter than a second. The light travel time across an emitting region must be smaller than the shortest observed variability timescale, or the change in brightness will be smoothed out by the variation in arrival time of photons emitted from different parts of the source. Thus the size of the source R must be less than the typical timescale over which the luminosity changes Δt times the speed of light, so $R \leq c\,\Delta t$. Thus a source that displays significant luminosity changes on timescales much shorter than a second, as the luminous X-ray sources are observed to do, must emit its light from a region smaller than a light-second across. Again, this is a size considerably smaller than most kinds of stars.

The general nature of these strong galactic X-ray sources is now well understood. They are double star systems consisting of a Sun-like main sequence or giant star and a much more compact object. The size of the orbit of this binary star system is such that material from the companion star can accrete onto the compact object. As described in chapter 2, such an accretion flow can generate strong high-energy emission. The accreting mass must be very compact to satisfy the size constraints imposed by the observations and to ensure that the gravitational potential well is deep enough to generate sufficient energy to power the source.

4.2 Varieties of X-Ray Binaries

While the observed phenomena from X-ray binaries are very diverse, it is clear that these systems are of two basic types. In the "low-mass" systems, the companion star is less massive than the accreting compact object. The gravitational force of the compact object deforms the companion star into a teardrop shape known as a *Roche lobe*. The point of the Roche lobe closest to the compact object is the inner Lagrangian point, where the gravitational forces of the two stars and the centrifugal force generated by the orbital motion balance precisely. Any material from the companion star that crosses the Lagrangian point will fall away from the companion and onto the compact object.

It is not surprising that systems of this kind generally have a mass ratio such that the accreting object is more massive than the star that is losing the mass. Angular momentum conservation ensures that mass transfer from a high-mass star to a lower-mass accreting object will bring the two objects closer together. If the mass transfer is driven by Roche lobe overflow, then the rate of mass transfer will increase if the objects move closer together, since this will make the Roche lobe smaller and increase the amount of overflow. This can lead to a runaway situation in which an initially small mass transfer rate can increase until enough of the mass-losing star is transferred that the mass ratio is reversed, and the mass-accreting star becomes the more massive of the pair.

The effect of mass transfer on the separation between orbiting objects can be determined by considering the total angular momentum of the two stars. For circular orbits, the total angular momentum J can be written as $J = m_1 V_1 D_1 + m_2 V_2 D_2$, where m, V, and D are the mass, velocity relative to the center of mass, and distance to the center of mass of the two objects, respectively. The velocity and distance of star 1 can be related to the total relative velocity between the two objects V_t as $V_1 = V_t(m_2/m_t)$, where m_t is the total mass of the two stars, $m_t = m_1 + m_2$. The distance of star 1 from the center of mass can be related to the distance between the two stars D_t by $D_1 = D_t(m_2/m_t)$. Similar relations apply to star 2. The relative velocity and distance of the two stars are related through Kepler's laws by $V^2 = (Gm_t/D)$. Combining all these relations gives

$$J = G^{1/2} m_t^{3/2} D_t^{1/2} (m_1 m_2/m_t^2).$$

If the total angular momentum J is conserved, and all the mass stays in the system—a situation known as *conservative mass transfer*—then J and m_t must remain constant. In this case, D_t must increase if $m_1 m_2$ decreases, and vice versa. The quantity $m_1 m_2$ is maximized when the two masses are equal, so mass transfer that brings the masses of the two stars closer together (that is, transfer from the more massive star to the less massive star) requires that D_t decrease. This condition leads to runaway mass transfer, while mass transfer from the less massive to the more massive star leads to a stable situation.

In contrast, the "high-mass" systems have companion stars that are generally more massive than the compact objects. Such high-mass stars eject large quantities of matter in the form of a stellar wind. A fraction of this wind can be captured and accreted by the compact object. Such mass transfer is not conservative, as most of the mass is lost to the system, carrying angular momentum with it, so the preceding derivation does not apply. In this case, a black hole can accrete matter from a higher-mass companion. These two kinds of systems are likely to have quite different evolutionary histories and are thus found in somewhat different circumstances. In particular, the high-mass systems tend to be found in regions of ongoing star formation—they must be quite young, since the lifetime of the massive companion stars is relatively short. In contrast, the low-mass systems can persist for billions of years, and thus have time to travel far from their birthplace.

4.3 X-Ray Accretion States

Observations of X-ray binaries continued to improve as increasingly capable X-ray observatories were launched. A key observation was that these sources are strongly variable, on timescales ranging from fractions of a second to many years. In particular, many sources proved to be *transients*, which rise from undetectable levels to become some of the brightest X-ray sources in the sky over a few days. These sources then gradually become fainter again, until after a few weeks or months they are again no longer observable. Recently, the most powerful X-ray observatories have detected X-rays from transients in quiescence, demonstrating that their X-ray flux rises by as much as eight orders of magnitude during an outburst. The quiescent state can persist for years to decades; indeed, some transients have been observed to go into outburst only once in the history of X-ray astronomy. In quiescence, the optical light from the companion star often dominates the flux associated with mass accretion, and observations of the companion provide crucial information about the source, as described in the next section.

Nontransient X-ray binaries also show very marked changes in behavior over short periods of time; the luminosity and the spectrum can change dramatically. Spectra of X-ray binaries are sometimes dominated by thermal emission of the kind expected from an accretion disk. Such spectra are superpositions of blackbody spectra, each at a different temperature, with higher temperatures being associated with the inner parts of an accretion disk. At other times, X-ray binaries display nonthermal spectra in the form of a power-law function of flux as a function of wavelength. The power law typically extends to much higher energies than the thermal spectrum. Much of the emission in these states comes from hot gas above and around the accretion disk. This gas, often referred to as an *accretion disk corona* (ADC) is thought to be the source of much of the nonthermal emission from X-ray binaries. The relative strength of the thermal and nonthermal components of the spectra can change dramatically within a single object. These changes are often associated with changes in overall brightness of the X-ray source, and with changes in the amount and kind of short-term variability. These changes are referred to as *state changes* and presumably result from changes in the rate and geometry of the flow of matter onto the compact objects.

X-ray binaries also display a variety of phenomena. The orbital period is often revealed by eclipses and other effects of changes in viewing angle during an orbit. Variability on shorter timescales is also generally seen, sometimes in the form of random "flickering" and sometimes as "quasi-periodic variations," in which a specific timescale is favored. A single object can display quasi-periodic variability on timescales much shorter than a second, orbital changes on timescales of hours or days, and state changes over many months.

As the state changes were studied it became clear that certain characteristic combinations of effects were often seen together, which led to many efforts to develop a taxonomy of *X-ray states* for X-ray binaries and to understand the transitions that are observed between them. It seems clear that the different X-ray states are associated with different kinds of accretion flows, such as disks or ADAFs. Two common states are the thermal state, in which most of the emission comes from an accretion disk, and the hard state, in which an outer disk is observable in optical and infrared wavelengths, while the inner accretion flow is from a nonthermal ADC and may also include a component from a jet or outflow (see figure 4.2). Observable radio emission, generally assumed to be from a jet, is often present in the hard state. The thermal state is generally associated with lower levels of variability than the hard state. A variety of other extreme and intermediate states have also been identified. The most appropriate definitions for the various states, the physics that gives rise to them, and the events that trigger transitions between states are active topics of current research.

4.4 Compact Objects

Not all the compact accreting objects in X-ray binaries are black holes. Other kinds of objects are sufficiently dense to generate X-rays by accretion. All of them are the natural result of stellar evolution—they are how stars end their life. Stars like the Sun are supported by gas pressure generated by the nuclear fusion process in their core; they are in a state of *hydrostatic equilibrium*, in which outward pressure forces balance the inward force of gravity. Eventually, however, a star exhausts its nuclear fuel, and hydrostatic equilibrium can no longer be maintained by gas pressure. The core of the star then collapses until some other source of pressure restores the equilibrium. Such pressure can be provided by the quantum mechanical effect of *Fermi pressure*, which arises because two particles are prohibited from occupying the same volume simultaneously. In stars like the Sun, the Fermi pressure of the electrons holds the star up when it reaches a density of approximately a ton per

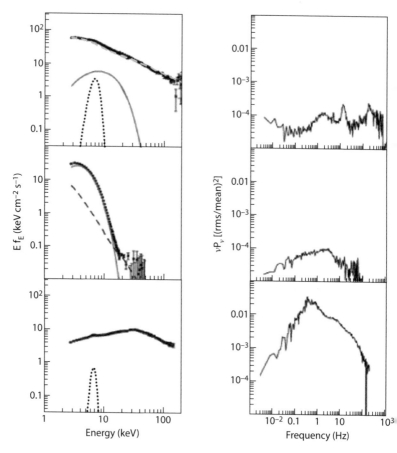

Figure 4.2. One current definition of X-ray states (from Remillard and McClintock 2006): the left-hand panel shows the spectrum, and the right-hand panel shows the temporal power spectrum (variability). Top panel: the steep power state; middle panel: the soft thermal state; bottom panel: the low hard state. The differences in spectral and temporal behavior between these states are clear. There are also a variety of intermediate states. From R. Remillard and J. McClintock, 2006, *Annual Review of Astronomy & Astrophysics*, 44: 49.

cubic centimeter, or a size comparable to that of the Earth. Such stars are known as *white dwarfs*. Mass-accreting white dwarfs are sometimes referred to as *cataclysmic variables* and are responsible for nova explosions and other phenomena. Some classes of cataclysmic variables emit X-rays but not in sufficient quantities to account for the strongest X-ray sources, like those identified by *Uhuru* in the galactic center.

However, not all stars end their life as a white dwarf. In the 1930s, the great astrophysicist Subrahmanyan Chandrasekhar showed that white dwarfs with a mass greater than $1.4\,M_\odot$ cannot support themselves against gravitational collapse—this bound is referred to as the *Chandrasekhar limit* after its discoverer. This result was controversial at the time—Sir Arthur Eddington famously remarked that "there should be a law of Nature to prevent stars from behaving in this absurd way."[1] One version of the derivation is given in section 4.8.

Stars that exceed the Chandrasekhar limit after their nuclear fuel is exhausted must continue to collapse—but to do so the electrons must be eliminated, which occurs through *electron capture* by the protons: $e + p \rightarrow n + \nu$, where ν is a neutrino. Essentially all the matter is transformed into neutrons, creating a *neutron star*. Neutron stars can be observed in the form of pulsars, in X-ray binaries, and in a few cases as isolated objects. The large number of neutrinos generated stream outward from the implosion. The enormous energy created by the implosion blows off much of the envelope of the star in a supernova

[1] *The Observatory*, 1935, 58 (no. 729): 38.

explosion. Neutron stars achieve enormous densities (millions of tons per cubic centimeter), comparable to the density of an atomic nucleus, and thus have extraordinarily deep gravitational potential wells. Accreting neutron stars could generate the X-rays observed from X-ray binaries, and indeed this is what many of the X-ray binaries are thought to be.

However, there is also an upper limit on the mass of neutron stars. They are held up by Fermi pressure of the neutrons, in much the same way as white dwarfs are held up by pressure from the electrons. Thus in principle there is an upper limit on the mass of a neutron star similar to the Chandrasekhar limit for white dwarfs. However, this limit is modified by the strong general relativistic effects present in neutron stars, which are only slightly bigger than their Schwarzschild radius. As shown in section 4.8, the mass-radius relationship for objects held up by Fermi pressure is inverted, in the sense that a larger mass results in a slight decrease in radius. Therefore, above some mass the radius of a neutron star will be less than its Schwarzschild radius, and the object will be a black hole.

The exact upper limit on the mass of a neutron star depends on the *equation of state* of neutron-dominated matter, that is, on the relationship between density and pressure. The correct equation of state to use for neutron stars is not completely understood—"soft" equations of state, which produce more compressible configurations, can result in an upper limit on the mass of a neutron star as low as $1.5 M_\odot$, while "hard" equations of state can result in neutron stars as massive as 2.2–$2.5 M_\odot$. The softest equations of state are now ruled out by observations of neutron stars with masses approaching $2\,M_\odot$, but the mass-radius relationship of neutron stars is still under study.

However, there are some basic requirements that any equation of state for neutron stars must satisfy. From the experimental point of view, equations of state must be compatible with the results of direct experimentation in nuclear physics, while from a theoretical perspective equations of state must be causal—that is, the speed of sound in the material must be less than the speed of light. This minimal requirement on the equation of state results in a maximum mass for a neutron star of around $3\,M_\odot$. Thus a compact object with mass greater than three solar masses is likely to be a black hole—it cannot be a white dwarf, since it is above the Chandrasekhar limit, and it cannot be a neutron star, since such an object would collapse to become a black hole.

In fact, there are ways in which a compact object with $M > 3\,M_\odot$ could avoid being a black hole. For example, strong differential rotation can hold up a neutron star as massive as six solar masses; however, the strong tangled magnetic fields inside a black hole would damp out such differential rotation in seconds. If the compact object is composed of exotic materials (e.g., particles containing strange or charm quarks), then the equation of state could be quite different from that of a neutron star, resulting in exotic objects quite different from standard neutrons stars. Finally, it is possible that general relativity requires modification in the limit of strong fields, in which case the concept of the Schwarzschild radius may not be applicable. The extraordinary nature of these alternatives reinforces the strong probability that a massive compact object must be a black hole.

4.5 Mass Measurements in X-Ray Binaries

Since compact objects with $M > 3M_{\odot}$ are almost certainly black holes, determining the mass of the accreting object in X-ray binaries becomes an important measurement. Fortunately, standard techniques of binary star astronomy can be used to make such mass determinations. In particular, Kepler's laws of orbital motion can be used to find the mass of orbiting objects from their velocity and orbital period. Repeated observations of the Doppler shift of an orbiting object can be used to construct a *velocity curve*, in which the radial velocity (that is, the speed toward or away from the observer) of the source is plotted against time. During an orbit, the object first moves toward the observer, then away, so the velocity curve is periodic at the orbital period. For circular orbits, the velocity curve is sinusoidal. From such a sinusoidal velocity curve one can compute the *mass function* $f = PK_*^3/(2\pi G)$, where P is the orbital period, and K_* is the amplitude of the sinusoidal velocity curve (see figure 4.3; note that despite its name, f is not really a function of mass but rather a quantity with units of mass). It is straightforward to show that

$$f = PK_*^3/(2\pi G) = M_x \sin^3 i / (1 + M_*/M_x)^2$$

where M_* is the mass of the observed star, M_x is the mass of the (unobserved) compact object whose accretion

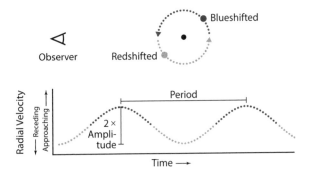

Figure 4.3. Radial velocity curve of X-ray binary with a circular orbit.

presumably generates the X-rays, and i is the angle of inclination of the binary orbit. The importance of this expression is that $f \leq M_x$. Thus if the velocity curve of the companion star can be measured, a minimum mass of the compact object immediately follows solely from considerations of the dynamics of the binary orbit. Such observations can be made if the companion star generates most of the optical emission from the system, which can occur in high-mass systems in which the companion star is intrinsically very bright, or in low-mass transient systems when the accretion flow is in quiescence. If $f > 3\,M_\odot$ this provides dynamical confirmation of a black hole in the system. There are now over a dozen "dynamically confirmed black hole candidates," that is, X-ray binaries in which the mass function f has been measured to be greater than $3\,M_\odot$.

But while it is true that $f > 3\,M_\odot$ implies $M_x > 3\,M_\odot$, it is not necessarily the case that $f < 3\,M_\odot \Rightarrow M_x < 3\,M_\odot$.

To determine the mass of the compact object from the mass function requires knowing the mass ratio M_*/M_x and the orbital inclination i. Thus two additional observational constraints are required beyond the mass function to fully determine the masses and inclination of the binary system. These constraints are generally provided by the rotational broadening of the stellar absorption lines, which is strongly sensitive to the mass ratio, and the amplitude of the ellipsoidal variations, which is strongly sensitive to the orbital inclination.

The mass ratio can be constrained by determining the rotational velocity of the surface of the companion star. For Roche lobe–filling systems, the companion star is tidally locked to the orbit, so the rotation period is always the same as the orbital period, as is the case for the Moon in its orbit around the Earth. If the rotational period is known, the rotational velocity at the equator depends on the size of the object, since $2\pi R/P_{rot} = V_{rot}$. The rotational velocity of the companion star can be measured by observing the breadth of the spectral lines from the star—the lines will be Doppler broadened because part of the rotating star is coming toward the observer and the other side is receding. Since the companion star fills its Roche lobe, its size is fixed by the semimajor axis of the orbit and the mass ratio, since these two parameters determine the size of the Roche lobe. The semimajor axis can be calculated from the known orbital period and the total mass, so measuring the rotational broadening gives a constraint on the total mass of the system and the mass ratio.

The orbital inclination can also be constrained by standard binary star techniques. In this case, the key observation is the change in brightness of the companion during the course of an orbit. Such an orbital light curve generally displays *ellipsoidal variations.* These variations in brightness are caused by the change in viewing angle of the companion star (see figure 4.4, top). The companion star is tidally distorted and hence not spherical. When viewed from the side, it displays a larger cross section, and thus appears brighter, than when viewed end on. Thus an orbital variation in brightness is expected in which there are two maxima of the light (when the star is viewed side on) and two minima (when the star is viewed from the front or the back). The amplitude of the variation depends strongly on the inclination, since in a face-on configuration the companion star always displays the same side to the observer, whereas an edge-on configuration maximizes the effect (see figure 4.4, bottom).

This variation is roughly similar to that of a rotating ellipsoid (hence the name "ellipsoidal" variations), but this is not precise, and it is now possible to make detailed computational models of the exact shape of the distended star. Currently used models now include both the effects of the unusual geometry of the star and possible contributions to the overall flux from the accretion flow. Excellent agreement between data and models can be found, implying that the inclination is well determined. There are now close to two dozen sources in which the mass of the compact object is well determined and significantly greater than $3\,M_\odot$.

Most of the known systems with $M_x > 3\,M_\odot$ are transient low-mass systems, in which the companion star can be observed when the accretion flow, and the luminosity

Ellipsoidal Variations

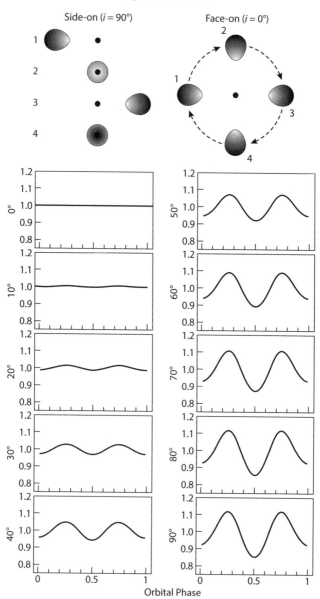

Figure 4.4.

associated with it, turns off. However there are also persistent high-mass systems with dynamical evidence for black holes. Within our own galaxy, only the famous system Cygnus X-1 falls into this category, but recently, dynamical evidence for such black holes has been discovered in systems in other galaxies. It should be remembered that the taxonomy of "high mass" versus "low mass" refers to the mass of the companion star, not that of the compact object. Interestingly, it appears that the most massive stellar black holes are found in the wind-fed "high-mass" systems. There also appears to be a significant lack of black holes in the 3–5 $M_\odot$ range. There is as yet no compelling explanation for the observed distribution of black hole masses in these systems.

4.6 Are High-Mass Compact Objects "Black Holes"?

Given the existence of a sample of "dynamically confirmed black hole candidates," that is, compact objects with measured masses above the 3 $M_\odot$ limit, it is important to determine whether these objects really do behave like the black holes predicted by general relativity. To date, the evidence suggests that they do. The most basic property of black holes is the absence of a surface, which has potentially observable consequences, as there are phenomena

---

← **Figure 4.4.** Ellipsoidal variations. Top panel: face-on and edge-on configurations; bottom panel: orbital light curves as a function of inclination.

associated with accreting gas that falls onto a surface. Such effects would not be expected in accreting black holes, although they might well occur in accreting neutron stars.

An example of such surface behavior are *X-ray bursts*.[2] When infalling gas lands on a neutron star, it must undergo a series of fusion reactions, eventually ending up as neutron-rich material that can be absorbed into the neutron star itself. Depending on the accretion rate, some of these thermonuclear reactions may occur explosively. Explosive nucleosynthesis occurs when the accreted material reaches a critical mass. Beyond this threshold the reaction occurs explosively in all the accreted material over a fraction of a second. Such explosions are observed in the form of bursts of X-rays, which can increase the luminosity of X-ray binaries by an order of magnitude over a small fraction of a second. X-ray bursts have been observed in many X-ray binaries; however, they have never been observed in any system in which the compact object mass has been shown to be above the $3\,M_\odot$ limit. The implication is that these more massive compact objects are indeed black holes and as such, lack surfaces on which the accreted material can accumulate. The observation of an X-ray burst from a massive compact object would constitute a significant challenge to the current belief that these objects must be black holes. But thus far, the extensive archive of X-ray bursts comes only from systems whose accreting objects have a mass consistent with neutron stars.

Another consequence of the existence of a surface on an accreting object is the existence of a *boundary layer* at

---

[2]There are actually several kinds of X-ray bursts. I refer here to the so-called "Type 1" events.

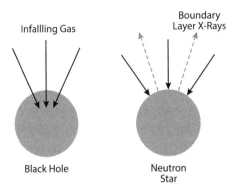

Figure 4.5. Accretion onto a black hole versus a neutron star. Accreting neutron stars emit boundary layer radiation, which black holes do not.

which the accretion flow terminates. This boundary layer is likely to generate significant amounts of radiation, as the kinetic energy of the infalling gas must be turned into some other form of energy at the point where the gas lands on the accreting object (see figure 4.5.) In the case of a classic accretion disk, in which the gas is in circular Keplerian orbits, the kinetic energy is always equal to half the potential energy, so half of the energy associated with the infalling material must be dissipated in the disk itself, and the other half must be associated with the termination of the disk at the accreting surface. Thus an accreting object with a surface would be expected to be twice as luminous as a black hole for a given mass accretion rate. However, it is hard to associate an observed spectrum with a mass accretion rate, so this test is hard to apply. In contrast, in some ADAF-like flows, almost all the energy is advected inward, and only a tiny fraction of the energy is dissipated during the infall. In these cases, accreting neutron stars would be expected to be orders of magnitude brighter than accreting black holes. In many cases the X-ray emission from low $\dot{m}$ sources (which are expected to have ADAF-like flows) appears to be higher for neutron stars than for dynamically confirmed black hole candidates, providing support for the idea that these objects may not have a surface.

4.7 Isolated Stellar-Mass Black Holes

The black holes in X-ray binaries represent only a small subset of the stellar-mass black holes that probably exist in our galaxy. Most stars with initial masses greater than $\approx 30\,M_\odot$ (the exact limit is only poorly determined) will likely retain enough matter to evolve into a black hole. While such massive stars are relatively rare, the galaxy contains around 10^{11} stars, so millions of black holes are expected. Only those that happen to be in binary systems with very specific configurations appear as X-ray binaries. The vast majority of black holes are not in such systems and thus cannot be detected in this way. So, it is natural to ask whether there might be other ways to identify black holes that are not contained in X-ray binaries. Two general schemes have been suggested, but both also depend on very unusual interactions with other objects, and successful identifications have been rare.

One approach is simply to look for black holes that are accreting mass from the diffuse interstellar medium (ISM) rather than from a companion star. An isolated black hole would be expected to accrete from the ISM according to the standard Bondi-Hoyle accretion process described in chapter 2. The mass accretion rate depends on the density of the ambient medium, which in general is very low. Thus the luminosity generated should also be very low. Indeed, the luminosity might be lower still if the accretion is in the form of an ADAF or other radiatively inefficient flow. So, such objects would generally not be expected to generate sufficient luminosity to be detected. However, the ISM is very inhomogeneous, and parts of the ISM are much denser than average. In particular, the molecular clouds in which star formation takes place have quite high density and might generate significant accretion if a black hole were to travel through them. However such dense clouds cover quite a small fraction of the galaxy, which greatly diminishes the number of such sources expected to be seen. Dense star-forming clouds also contain young stars that might generate X-rays and other high-energy radiation in other ways. The combination of these factors means that no isolated black hole has compellingly been identified in this way.

Another technique for identifying isolated black holes is *gravitational microlensing*. Gravitational lensing occurs when a massive object is directly in the line of sight of a luminous background object. The massive object acts much like an optical lens and bends the light rays from the background source, resulting in a magnification of the light of the background object. Such lensing can occur only when the two objects are very precisely aligned. Since objects in the galaxy are in constant motion, this magnification is not expected to persist; rather, the magnification is expected to grow over the course of a few weeks and then to decay again. Thus a black hole passing in front of a background star might magnify the light of the star by an observable amount for a few weeks, even if the black hole itself emits no radiation at all.

A chance alignment precise enough to produce significant magnification of the background object is quite unlikely. Therefore, millions of background stars must be observed on a daily basis to provide a significant event rate. Several large survey programs have been carried out to do this, focusing on the Galactic Bulge and the Large Magellanic Cloud, fields in which many background stars can be observed simultaneously.[3] The original goal was to determine whether the "dark matter" in the galaxy required by galaxy rotation curves and other evidence could be explained by dim stars or compact objects. Such objects might be unobservable directly but might reveal their presence by generating a significant number of microlensing events. The observed event rates demonstrated that such macroscopic objects could not, in fact, account for the dark matter, and relatively few events were observed that were compatible with massive ($M > 3\,M_\odot$) nonluminous lenses. But while stellar mass black holes clearly cannot account for the dark matter, the microlensing results are nevertheless compatible with a large number of isolated black holes in the galaxy.

---

[3] See C. Alcock et al., 2001, *Nature* 414:617; and A. Udalski et al., 1997, *Acta Astronomica* 47:319.

4.8 The Chandrasekhar Limit

The structure of a star is determined by the balance
between gravitational forces, which pull material inward
toward the center, and pressure forces pointed outward. In
stable stars, the forces balance at every point, so that the
star neither implodes nor explodes and is thus in a state of
hydrostatic equilibrium. For spherically symmetric stars, the
gravitational acceleration at any point at distance R from
the center is due only to the material located inside that
point and is equal to $a_{\mathrm{grav}} = -GM(R)/R^2$, where $M(R)$
is the mass contained between the center of the star and the
distance R. The minus sign denotes that the acceleration
is directed inward. The pressure depends on the nature of
the material in the star. In stars like the Sun, the pressure is
gas pressure and is determined by an equation of state such
as the ideal gas law $P \propto \rho T$, where ρ is the density and
T is the temperature. However, the pressure thus defined
is exerted in all directions, whereas what is required is the
net pressure outward. This is determined by the pressure
gradient, that is, the difference between the pressure just
inside a given point R from the center and the pressure
just outside that point. Thus in general, a star must satisfy
$(1/\rho)dP/dR = -GM(R)/R^2$.

One of the features of an equation of state like the
ideal gas law is that the pressure is directly proportional
to the density. The gravitational force inward is also
related to the density, in that $M(R)$ is determined by
integrating the density from the center of the star out to R.
If the gas pressure is linearly dependent on the density,
as would be the case if the temperature were constant

(an *isothermal* configuration), then it can be shown that hydrostatic equilibrium cannot be achieved—increasing dP/dR requires the density to rise quickly toward the center of the star, which creates a large value of $M(R)$, and the pressure forces cannot keep up. Therefore gas pressure can balance gravitational force only if the temperature also rises toward the center of the star, providing another contribution to dP/dR. But if there is a temperature gradient, basic thermodynamics requires that heat flow outward from the hot interior. If there is a source of energy in the center of the star, the heat flowing outward can be replaced, so that the temperature gradient and hydrostatic equilibrium can be maintained. But without a central energy source, the star evolves toward an isothermal situation, gravity overcomes the gas pressure, and the central regions of the star collapses.

This collapse continues until the gas in the star becomes *degenerate*. Gas is degenerate when the Pauli-Fermi exclusion principle, which states that two fermions cannot be in the same place at the same time, becomes an important factor. When the gas becomes sufficiently dense, the distance between the electrons becomes sufficiently small that the exclusion principle creates a *Fermi pressure* that keeps the electrons from coming too close to one another. This Fermi pressure increases as a high power of the density and thus is sufficient to balance the inward gravitational force even for an isothermal star. White dwarfs are stars that have exhausted their nuclear fuel and contracted to such a high density that Fermi pressure from the electrons suffices to maintain hydrostatic equilibrium.

In a nonrelativistic gas (that is, a gas in which the energy arising from the momentum of the particles is much less than that associated with the rest mass of the particles) an approximate radius for a star held up by degeneracy pressure can be derived as follows. If a star contains N fermions and has radius R, then a typical separation x between individual fermions will be $x \approx (V/N)^{1/3} \approx R/N^{1/3}$, where $V \propto R^3$ is the volume of the star. The momentum of each particle must be $1/x$, and the average Fermi energy per particle must be $E \approx p^2/m$, where m is the mass of the particle. In an equilibrium state, the typical gravitational energy of each particle $\approx GM(R)m/R$ must balance the Fermi energy; since $M(R) \approx Nm$, we can equate the Fermi energy and the negative gravitational potential energy per particle and find that

$$R \approx 1/(Gm^3N)^{1/3}.$$

This equation has the curious result that, unlike ordinary stars, massive degenerate stars (with larger values of N) are *smaller* than their lower-mass counterparts. In essence, the greater gravitational force generated by a more massive star must be countered by greater Fermi pressure, which requires that the particles be closer together and the total configuration be smaller. As the electrons are packed in closer together, x, the typical distance between them decreases, and the Fermi momentum increases. Eventually, the Fermi energy becomes comparable to the rest mass energy of the electrons, and the gas becomes relativistic. For relativistic gas the energy per particle is $N^{1/3}/R$. This means that the gravitational energy and the Fermi energy have the same dependence on R, which drops out of the equilibrium equation. So instead of a relationship between N and R, as is found for a nonrelativistic gas, we find that N depends only on constants:

$$N \approx (1/Gm^2)^{3/2}.$$

Thus there can be only one value for N, and for a given mix of atomic species, only one mass, for a fully relativistic gas. This mass is the Chandrasekhar limit—greater masses cannot be sustained by Fermi pressure against gravitational collapse. Typical white dwarfs contain no hydrogen or helium, since those are the elements that provide energy from nuclear fusion to maintain gas pressure in nondegenerate stars—it is only when the hydrogen and helium are exhausted that a white dwarf forms in the first place. This means that white dwarfs are typically dominated by elements like carbon, nitrogen, oxygen, neon, and magnesium, for which there are two nucleons per electron. In this case the Chandrasekhar limit turns out to be $1.44\ M_\odot$.

We might then ask, what happens to a degenerate configuration that happens to be greater than the Chandrasekhar masses. Gravitational forces dominate, making the star contract and thus forcing the electrons closer together than is allowed by the uncertainty principle, so the star must lose the electrons altogether. Electrons and protons combine to make neutrons and neutrinos, in the process of inverse beta decay. Thus the entire mass of the star is turned into neutrons, forming a neutron star, and huge numbers of neutrinos are released. This transformation is observed in the form of supernovae, which often leave behind neutron stars. Neutrons are also fermions, so in principle the same ideas apply: nonrelativistic configurations get smaller with increasing mass until the particles become relativistic, at which point a limit similar to the Chandrasekhar limit applies, and the star collapses. The main difference is that because neutrons are more massive than electrons, the equivalent Fermi energy is generated by much smaller distances between them, and thus neutron stars are much smaller and denser than white dwarfs. In fact, however, general relativistic effects come into play before the equivalent of the Chandrasekhar limit becomes relevant. When the mass of a neutron star becomes greater than 2.2–3 $M_\odot$ (the exact value depends on the specifics of the neutron star equation of state), it collapses to become a black hole.

This derivation of the Chandrasekhar limit is approximate, as it assumes a constant density and thus a constant separation between the fermions. Real degenerate stars have a density gradient, and thus the requirement of hydrostatic equiliibrium must be satisfied at every radius, which results in differential equations for mass, density, and pressure as a function of radius that must be solved. This can be done with results comparable to those described here.[4]

---

[4]See, for example, S. Shapiro and S. Teukolsky, *Black Holes, White Dwarfs, and Neutron Stars: The Physics of Compact Objects* (New York: Wiley, 1983).

5

SUPERMASSIVE BLACK HOLES

Stellar-mass black holes are clearly common consequences of stellar evolution, but they are not the only kinds of black holes identified by astronomers. Much more massive black holes are located in the center of many, and perhaps all, galaxies, including our own. These black holes are referred to as *supermassive black holes*, sometimes abbreviated "SMBHs." They are responsible for a range of phenomena originating from objects described as *active galactic nuclei* (abbreviated AGN), which were first observed in the form of *quasi-stellar objects* (QSOs) or *quasars*.[1] AGN are among the most luminous objects in the Universe and can be observed at great distances. The distances can be so great that the light travel time from the AGN to Earth is a large fraction of the age of the Universe—we are seeing AGN not as they are today but as they were when the Universe was much younger. They are therefore often used to probe the evolution of the Universe.

---

[1] The terms *QSO* and *quasar* are sometimes used interchangeably, but sometimes QSO is used to denote objects discovered by their optical properties, whereas quasars denote those discovered in radio frequencies.

5.1 Discovery of Quasars

QSOs were identified in the 1960s as a class of very blue objects that appear as single points of light, like stars, in contrast with galaxies and nebulae, which are extended objects. However, their spectra are quite different from those of ordinary stars. In particular, the line emission from the QSOs did not correspond to any known atomic transitions. The key to understanding QSOs was the realization that the spectral features could be identified provided the objects were strongly redshifted, with redshifts as high as $\Delta\lambda/\lambda \approx 1$. This result implied that QSOs are extragalactic objects located at cosmologically large distances whose redshifts are caused by the expansion of the Universe. Such large distances required QSOs to be the most distant and most intrinsically bright objects yet discovered.

The cosmological distance scale for QSOs was not immediately accepted by the full astronomical community. At the time, there was still support for the "steady state theory" of cosmology, which requires continuous matter creation throughout the cosmos to preserve a constant density as the Universe expands. It was suggested that QSOs might be the site of this matter creation. Soon, however, the evidence for the Big Bang and for cosmological distances for QSOs became overwhelming. But for some time the cosmological distance scale for QSOs was disputed, primarily on the grounds that high-redshift quasars were often seen close to low-redshift galaxies. If the galaxies and QSOs were, in fact, part of the same object, then the high QSO redshifts could not be due to their distance. A few prominent astronomers, most notably Halton Arp, still question the cosmological distances for QSOs. But the evidence for associations between QSOs and low-redshift objects have become very much weaker, and holdouts like Arp are no longer taken seriously. The spatial coincidences are now interpreted as chance superpositions. A crucial effect is that distant QSOs can be gravitationally lensed by nearby galaxies, which makes background quasars brighter and thus easier to detect than they otherwise would be. When this effect is taken into account the number of chance superpositions is in accord with expectations. There is thus no reason to doubt the distance scale of QSOs, and the associated high luminosities.

In fact, the QSO redshifts provided a key piece of evidence in favor of Big Bang cosmology. As surveys for this new class of objects began to be carried out it became apparent that there were more QSOs at high redshifts than expected by extrapolating the number of low-redshift QSOs, which suggested that the population of QSOs has changed over cosmic time. QSOs with high redshifts, and thus large distances, are seen as they were in an earlier epoch of the Universe, since the light travel time is a significant fraction of the age of the Universe. Thus there were apparently more QSOs at earlier times than there are now. Such changes in the overall distribution of cosmic objects are expected in a Big Bang cosmology, in which there may be changes in the bulk properties of the Universe such as density and age, but not in the steady state theory, which as its name implies, requires that the cosmos not change its overall nature.

It soon became clear that QSOs are closely related to the previously discovered class of *Seyfert galaxies*, the centers of which exhibit quasar-like activity. As observational techniques improved, QSOs and quasars were firmly identified as an extreme case of the general phenomena of AGN, extreme in the sense that they are unusually luminous and thus can be seen at distances great enough that the host galaxy itself is invisible or difficult to detect.

Two basic features of AGN provide strong support for the idea that these sources are powered by accretion onto very massive black holes. First, significant variability is observed on timescales as short as a few days, and in some cases less than a day. This limits the size of the emitting region to a maximum of ≈ 1 light-day or $\approx 3 \times 10^{13}$ m. At the same time, the cosmological distances imply luminosities as high as 10^{44-47} ergs^{-1}. To avoid outshining the Eddington limit by orders of magnitude requires a central mass $M_c \gg 10^5 \, M_\odot$. For extreme cases, much of the emission must come from within a few hundred Schwarzschild radii of the central mass. It would be difficult to produce so much energy in a small region without involving black holes in some way.

More support for the idea that AGN must be powered by accretion onto black holes comes from their overall energy distribution and spectra, which are remarkably similar to what would be expected if the X-ray binaries were scaled up in mass by many orders of magnitude. In particular, one of the key observational characteristics of QSOs is their blue color, and space-based observations, particularly those conducted with the *International Ultraviolet Explorer* satellite, confirmed that many AGN have high levels of ultraviolet radiation as well. Thermal emission from accreting black holes should have peak at a wavelength λ_{max} that scales with mass as $\lambda_{max} \propto M^{-1/4}$. Thus the ultraviolet peak in thermal radiation shown by the QSOs, sometimes called the "Big Blue Bump," is analogous to the thermal X-rays seen in the X-ray binaries.

5.2 Active Galaxies and Unification

QSOs and quasars are related to a variety of categories of unusual galaxies. Many QSOs proved to be radio sources and thus seemed to be related to the class of *radio galaxies* which had already been identified as having extended radio emission. As noted earlier, the Seyfert galaxies, identified by the strong optical emission lines emanating from the central regions of the galaxy, appeared to be QSOs situated within a host galaxy. As these phenomena were studied in more detail it became apparent that the central regions of galaxies showed a very wide range of phenomena, sometimes referred to as the "AGN Zoo." A bewildering array of classification schemes was developed, each starting with one or several of the various observed characteristics described next.

Radio intensity. The early observations of quasars (identified as radio sources) and QSOs (identified as optical sources) had already demonstrated that the ratio of radio emission to that in other wavebands varies widely. As more details of AGN began to be understood, it became clear that about 10% of AGN are "radio loud," with significantly more radio flux than their radio-quiet counterparts.

Radio morphology. Radio galaxies had been identified even before quasars were discovered. Detailed maps of the bright radio emission from these sources show that the radio emission often emanates from two "lobes" surrounding the center of the galaxy. Some of these lobes extend for hundreds of kiloparsecs from the center of the galaxy. In some cases the center of the galaxy also emits observable radio flux, and "bridges" can be identified connecting the central source to the lobes. It is now generally believed that the substantial amount of energy required to power the radio lobes is generated in the galactic nucleus and transported to the lobes through collimated jets of the kind described in chapter 3.

Optical spectra. Many AGN show emission lines in optical and infrared wavelengths, and was the high redshifts of these emission lines that led to the identification of the QSOs as objects at cosmological distances. But the structure of these emission lines varies significantly among AGN. Some AGN have "broad lines," with widths corresponding to Doppler motions of thousands of kilometers per second. The breadth of these emission lines is generally interpreted as being due to a large velocity dispersion of the gas that produces the lines. The range of velocities of the emitting gas results in changes in the Doppler shift of the lines away from the mean redshift of the AGN. Superposing emission from gas with different velocities results in a broadened line. In addition having broad lines, most AGN also have much narrower lines. There are narrow-line objects that do not exhibit broad lines, but the reverse is not generally seen, although there are some AGN (known as *BL Lac objects*) in which line emission is not observed at all.

Polarization. The radiation observed from AGN is often strongly polarized, which requires that the radiation source be nonthermal (in contrast with starlight) or that the radiation pass through or be reflected by some kind of polarizing medium.

Host galaxy characteristics. The original distinction between QSOs and Seyfert galaxies rested on whether the host galaxy could be identified and thus on the relative luminosity of the galaxy and the nucleus. Seyferts were originally identified as a subcategory of spiral galaxies, but AGN can occur in galaxies of many shapes and sizes.

Time variability. AGN are strongly variable objects. Some AGN are known to vary in luminosity by as much as a factor of 2 in less than a day. More commonly, the timescales of variability are longer, but some level of variability is observed in almost all cases.

This wide range of phenomena led to attempts to impose order on the AGN Zoo through some kind of unifying principle. The most successful approach to *AGN unification* was the suggestion that much of the diversity of AGN phenomena could be explained simply by viewing angle.[2] As seen in figure 5.1, a generic AGN is now thought to comprise a number of components,

---

[2] A classic review of AGN unification is provided in C. M. Urry and P. Padovani, 1995, *Publications of the Astronomical Society of the Pacific* 107:803.

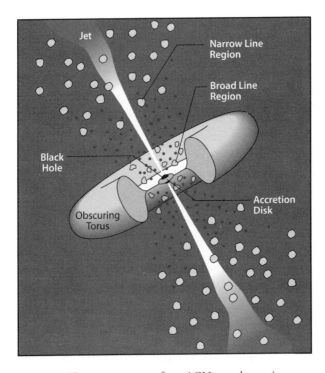

Figure 5.1. The components of an AGN are shown in cartoon form. An observer looking down onto the jet sees a blazar. Looking from the side, observers see a variety of AGN, depending on how much of the central region is obscured by the torus. After C. M. Urry and P. Padovani, 1995, *Pub. Astron. Soc. Pacific*, 107: 803.

in particular, the following:

- The central source. At the center of the galaxy is a black hole, of a million to a few billion solar masses. The vast energy emitted by AGN originates in accretion onto this black hole.

- The accretion disk. Just outside the black hole is an accretion disk, which feeds the black hole with material, much like in stellar mass systems. The source of the accreting material is not clear—it could be interstellar gas funneled down toward the center of the galaxy, or the remnants of stars that have ventured too close to the black hole and been torn apart by tidal forces. Presumably, the rate of gas accretion can vary strongly from source to source, leading to a wide range of luminosities, including the possibility of central black holes in which too little mass is being accreted for an AGN to be observed at all.

- The "broad-line region." Outside and above the accretion disk are clouds of gas which are ionized by radiation from the accretion disk. As the electrons recombine they emit observable spectral lines. Since these clouds are relatively close to the black hole, they travel at high velocities, and thus the emission from the lines is Doppler shifted from the mean value for the AGN as a whole. The combined emission from many of these clouds as they move in different directions then creates broad emission lines, which are seen in many AGN.

- The torus. Outside the broad-line region is a dusty opaque torus, which obscures material within it but radiates in the infrared. This torus is created by the outer edges of the accretion disk, where the temperature is low enough for dust to form, and the disk itself fragments into clumps. Such a region is not found in X-ray binaries, because it would lie outside the Roche lobe of the accreting object, and the angular momentum and energy in the outer disk is transferred by tidal forces back into the binary orbit. In AGN viewed from the side (sometimes called "Type 2 AGN"), the torus prevents the broad-line clouds from being seen directly, thus accounting for explaining the existence of AGN without broad lines. Type 1 AGN, in contrast, are viewed from the top, so their inner regions are not obscured by the torus. In some AGN, the broad lines can be seen only in polarized light, which suggests that light from the obscured broad-line region is being reradiated toward the observer's line of sight outside the torus.

- The "narrow-line region." Beyond and above the torus are more gas clouds, which again can be ionized by radiation coming from the heart of the AGN. These clouds also emit spectral lines, but since they are farther from the central source, they move more slowly, so the lines are much less Doppler broadened. In some nearby AGN, the narrow-line region can be seen to extend well beyond the central regions of the galaxy.
- The jet. As was discussed in chapter 3, some fraction of the accreting material is redirected into a jet moving outward perpendicularly to the accretion disk at close to the speed of light. These jets can create dramatic observational phenomena. In cases when the jets are pointed toward an observer, strong relativistic boosting can

dramatically change the observed characteristics of AGN. Such objects are called *blazars*.

- Radio lobes. The jets can transport large amounts of energy far away from the nucleus. Eventually, this energy is deposited into the galactic or inter-galactic gas. This energy can then power the radio lobes that are observed far from the galactic nucleus.

Thus the wide range of AGN activity is ultimately powered by accretion onto an SMBH, whose mass can be as high as several billion solar masses. This does not mean that all aspects of AGN phenomenology are understood—in particular, the circumstances that distinguish between radio-loud and radio-quiet sources are not explained by simple unification models, and many aspects of the detailed physics of AGN are still poorly understood. But the basic characteristics described here seem quite solid, and essentially all current work on AGN is being carried out within this framework.

5.3 Superluminal Jets and Blazars

Jets can be observed as collimated streams of material extending outward from AGN, frequently terminating in radio lobes. Often, the jets appear to bend, implying that the direction of jet emission precesses over time. The jets are often lumpy or knotty, suggesting that the jets turn on and off in specific jet events. They carry considerable amounts of energy and in some cases may dramatically influence the interstellar medium throughout the galaxy. A feedback loop is generated as the jet energy derives from accretion from the galaxy and in turn influences the galaxy, heating the interstellar gas and potentially changing the accretion process. Another key feedback process is the formation of new stars in gas that is gravitationally influenced by the SMBH. These AGN feedback processes can be difficult to untangle and are the subject of considerable current research.

As described in chapter 3, in some cases jets and knots are observed to travel across the sky at speeds apparently greater than the speed of light. Such superluminal motion is an illusion but requires the actual space velocity of the jet to be highly relativistic. In cases in which the jet is pointed almost directly toward the observer the viewing angle results in the blazar phenomenon (see figure 5.2). As noted in chapter 3, relativistic effects boost the intensity and energy of the jets and make the time variability shorter. In the extreme case of blazars, radiation from the jet can dominate all the other sources of radiation. Blazars thus provide excellent labs for studying jet phenomena, since there is effectively no contaminating radiation from other parts of the accretion flow.

Blazars are bright at all wavelengths, with synchrotron emission in the radio through optical, comptonized radiation at higher energies, and, finally, the most energetic observable photons. Orbiting satellites, including the recently launched *Fermi* satellite, show that blazars are among the brightest objects in the sky at photon energies above 10 GeV, and ground-based air shower arrays have identified photons of tera-electron-volt energies coming

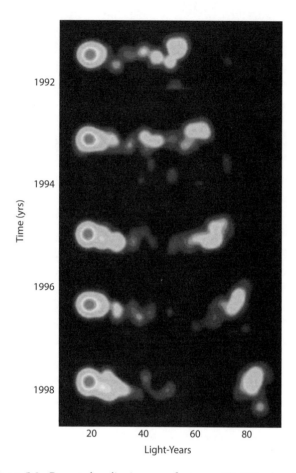

Figure 5.2. Repeated radio images of quasar 3C 279, showing superluminal motion. The radio lobe on the right appears to have moved almost 30 light-years between 1992 and 1998, which implies faster-than-light speed. This is in fact an optical illusion, as explained in chapter 3. Image courtesy of NRAO/AUI.

from these sources. This high-energy radiation is thought to come from Compton upscattering of the synchrotron emission generally observed in radio and infrared wavelengths. This high-energy radiation is then observed to be even brighter and more energetic due to the Doppler boosting provided by the motion of the jet.

Blazars also often vary substantially on quite short timescales—indeed, one subclass of AGN, the *optically violently variable* (OVV) sources, were identified from this characteristic. The OVVs are now understood to represent a particular observational characteristic of blazars. Again, this strong observed variability is a combination of intrinsic properties of the jets and the relativistic boosting from the jet motion. Jets are clearly unstable, both in space and in time. The events that launch the jets are highly episodic, so that specific "jet events" can often be identified as sudden luminosity flares. In blazars, the time variability of the jets is enhanced by the relativistic time compression, so that strong changes in luminosity occur on timescales of a day or less.

One difficulty in understanding blazars is that the same Doppler boosting that makes the jet so visible has the effect of making the other components of the AGN harder to see. The presence of the accretion disk is much less apparent in blazars than in other AGN, and many blazars have no observable line emission at all (the BL Lac objects described in section 5.2). Blazars thus provide wonderful laboratories for studying the propagation and emission from jets, but the connection to the underlying accretion processes that presumably provide the ultimate source of energy is not always clear.

5.4 Nonaccreting Central Black Holes

Given the large observed variability in luminosity observed from AGN, the question naturally arises whether galaxies that are not observed to have AGN may nevertheless contain central black holes which are simply accreting too little material to be observable. Such a non-AGN black hole has been identified most dramatically in the center of our own Milky Way. The galactic center is generally taken to be coincident with the radio source Sgr A*, which has also been identified as an X-ray source whose characteristics are generally similar to what would be expected from a scaled-up quiescent X-ray binary. There are severe difficulties in observing Sgr A* in optical wavelengths, since dust near the galactic center blocks the optical light from the source. However, the region can be observed in the infrared, which allows stars near Sgr A* to be identified.

By reobserving the region for many years, it has been possible to follow the motion of the innermost stars as they orbit around Sgr A*. These motions are small, and the observations require adaptive optics techniques that compensate for the blurring effect of the Earth's atmosphere. But such techniques have been available for over a decade, comparable to the orbital period of some of the stars. Since the distance to the galactic center is known, the spatial scale of these orbits is known, as are the orbital periods, so the mass of the central object can be computed. This turns out to be around 4 million solar masses, confined to a region comparable in size to our solar system. The total energy emitted by Sgr A* is less than that of the most luminous stars, so the mass-to-light ratio of this central source is enormously high. The combination of mass, compactness, and low luminosity leaves little doubt that this central source is a black hole that is currently in a very low accretion state. Models of the spectrum of Sgr A* suggest that the flow is in the form of an ADAF, as is apparently the case in most quiescent binary systems.

Such observations are difficult to make in other galaxies. But while individual stars are hard to observe in the center of distant galaxies, the distribution of light can be measured. In many cases, this light forms a cusp rising sharply toward the center, suggesting the presence of a large central mass. In some cases, the width of spectral lines reveals the spread of velocities of the stars contributing to the light. Often, the velocity dispersion also rises sharply toward the center, again implying the presence of a large unseen central mass. So, it seems likely that many, perhaps all, non-AGN galaxies contain central black holes as well.

5.5 Mass Determinations for Extragalactic SMBHs

The mass of the black hole in the center of our galaxy can be determined directly by the motions of individual stars. However, black holes in the nuclei of other galaxies cannot be observed in this way, since the distances are too great for stars close to the central object to be observed individually. However, the masses of some nearby super-massive black holes can be determined by other means. One of the most dramatic early results of the Hubble Space

Telescope (HST) was an observation of the gas around the supermassive object in the nucleus of the nearby giant elliptical galaxy M87. The high spatial precision of HST allowed measurements to be made of the Doppler shift of gas on either side of the galactic center. The results showed that the gas is orbiting the central object with a velocity of 550 km s^{-1} at a typical distance of about 20 pc (6×10^{17} m) from the galactic center. Using the standard orbital velocity formula $V = \sqrt{GM/R}$, the mass of the central object was determined to be at least 4 billion times the mass of the Sun, about a thousand times greater than the black hole in our own galaxy.

This observation did not in itself prove that the central object must be a black hole—the Schwarzschild radius of the black hole is only around 10^{13} m, orders of magnitude smaller than the observed orbits of the gas. Conceivably, a very dense cluster of nonradiating objects that are not themselves black holes might provide the mass required for the observed orbits of the gas. But the precision of the HST measurements reduced the maximum size of the region containing the central mass by a large factor. This meant that any such cluster must have a very high density of objects, which would undergo relatively close encounters regularly. The trajectories of the objects would be significantly altered by one another during the course of a *relaxation time* T_r, which can be shown to be of order $T_r = \sigma^3/(G^2 M^2 N)$, where σ is the velocity dispersion of the cluster, M is the mass of a typical star, and N is the number density of the stars. As the cluster relaxes, the density profile of the cluster changes such that the density at the core increases, and energy is removed from the cluster by stars that escape from it altogether. After a large number of relaxation times, the core density becomes very high.

In the case of a putative cluster of compact objects in the center of M87, the large density implied by the size limits on the cluster required by the HST results requires a relaxation time much shorter than the age of the galaxy. One would thus expect the cluster to have collapsed, possibly generating a black hole but certainly lacking the stability required to persist for any substantial length of time. If the relaxation time is short, as must be the case in the central regions of M87, no conceivable cluster of starlike objects would be able to survive the large number of collisions that would inevitably occur. Thus even if the central mass were contained in dim stars, those stars would collide and collapse into a single dense object within a short time.

A similar geometric approach takes advantage of the existence of *water masers* in the inner parts of the accretion disks around central black holes in AGN. Masers are the microwave equivalents of lasers and generate strong, sharp emission lines that can be observed with radio telescopes. Long-baseline interferometry using widely separated radio antennas, is used to obtain very high spatial precision— distinct masers are observable at separations of only a few milliarcseconds from the center of the galaxy, corresponding to linear separations of a fraction of a parsec. The redshift and blueshift of individual masers allows compelling Keplerian velocity curves to be obtained, leading to very precise measurements of the central black hole mass. That the individual maser measurements trace out a $R^{-1/2}$ velocity curve indicates that the mass is central. The small volume enclosed by the masers also supports the idea that all the mass being measured comes from a single black hole. In some cases the masers are seen to physically move across the face of the galaxy over the course of years. Equating the angular motion across the sky to the velocity measured from the Doppler shifts permits an accurate distance measurement also to be obtained. There are by now over a dozen maser measurements of black hole masses in galaxies. The exceptionally precise positional accuracy possible with maser measurements has proved to be particularly useful for measuring galaxies with moderate ($\approx 10^7 \, M_\odot$) mass AGN.

While these direct dynamical measurements of black holes are generally very reliable, they can be carried out on only a small handful of AGN. Even moderately distant AGN have angular scales on the sky that preclude separating the inner parts of the disk from the center of the galaxy. And some indication of the inclination of the disk is necessary to determine an accurate mass—in the case of the masers, the disk must be oriented almost edge on. Thus for most AGN, other less direct methods must be used.

One fruitful approach is known as *reverberation mapping*. This technique uses the emission features in the broad-line region that originate when gas in the clouds is ionized by radiation from the central source. An increase in the luminosity of the central source thus results in an increase in the line emission. But there is a time delay between the observation of an increase in the ionizing radiation and the corresponding increase in line emission, which arises because radiation travels directly from the central source to the observer, whereas the ionizing radiation must in general travel across the line of sight to get to the gas clouds of the broad-line region. Therefore the light travel time from the central source to the broad-line region to the observer is longer than the light travel time from the central source directly to the observer. Measuring this delay allows us, in principle, to determine the size of the broad-line region. But the width of the emission lines reveals the velocities of the broad-line region clouds. If the clouds are gravitationally bound to the central mass, the combination of distance and velocity can once again be used to determine its mass.

The observations required for good reverberation mapping measurements are demanding: the time delays are typically weeks to months, so the central source and the line emission must be observed over long periods of time. An element of luck is also required, in that there must be a readily observable change in luminosity of the source that generates an unambiguous response in the lines. The ideal is a single large change in luminosity that is sustained over a time that is longer than the light travel time across the broad-line region, so that the response of the whole observable line can be observed. Changes that are shorter or smaller become harder to observe, and disentangling multiple or continuous luminosity changes can be difficult. Nevertheless, a few dozen examples of mass measurements from reverberation mapping exist.

These mass measurement reveal a correlation between luminosity and size of the broad-line region. This allows the luminosity of the AGN, which is relatively easy to determine, to be used as a proxy for the size scale in systems for which the appropriate long variability studies required for reverberation mapping have not been carried out. Large surveys, in particular the Sloan Digital Sky Survey, containing many thousands of AGN spectra are available. Large numbers of AGN masses can be determined by converting the observed luminosity to a rough size and then using the width of the lines as a measurement of velocity. When this is done, a remarkable feature of AGN masses emerges, namely, a close relationship between the mass of the black hole buried at the center of the galaxy and that of the galaxy itself, often referred to as the *M–σ relationship*, where sigma is the velocity dispersion of the galaxy, and thus provides a measurement of the galaxy mass.

This relationship came as a surprise, since it requires a close bond between the origin and evolution of the central black hole, whose gravitational influence is quite local, and that of the host galaxy. One hypothesis for this bond is *hierarchical growth* of galaxies. This model of galaxy evolution postulates that protogalaxies are relatively small, contain small black holes, and grow by successive mergers. When these mergers occur, the black holes at the centers of the merging galaxies fall to the middle of the new, bigger galaxy and themselves merge. Thus studies of the evolution of supermassive black holes and of galaxy evolution appear to be intertwined.

This somewhat unexpected juxtaposition of two areas of astrophysics is currently prompting intense research. The evolution of galaxies is now being studied directly by comparing observations of high-redshift galaxies, which are observed as they were long ago due to the light travel time from these distant systems, with observations of nearer systems. Similarly, temporal changes in AGN-related phenomena can be studied by examining changes in the demographics of AGN with redshift. It is now possible to carry out detailed computer simulations of the merger process for galaxies and the black holes they contain and compare the results with the observations. And the black hole mergers themselves may soon be observable as sources of gravitational wave radiation, as described in chapter 9.

6

FORMATION AND EVOLUTION OF BLACK HOLES

Most objects in the Universe are not black holes. One might wonder why not: any collection of matter with negative total energy (including the intrinsically negative gravitational potential energy) tends to collapse due to its self-gravity. This tendency, if unchecked, will pull all matter together until it falls within an event horizon. So why shouldn't any gravitationally bound object quickly become a black hole? Alternatively, if the Universe is expanding faster than its own escape velocity, and the total energy of the Universe is positive, one might expect all matter to diverge into an increasingly sparse plasma and form no objects at all. Thus there appear to be two possible configurations for the Universe, one in which it has negative total energy and quickly collapses into one or a few black holes, and one in which it has positive total energy and expands into an infinitely large, sparse, cold, unfilled void. Obviously, our Universe has managed to avoid these two extremes and is sufficiently close to zero total energy that a variety of condensed objects can form even within an overall expansion.

It is something of a mystery why the initial conditions of the Universe landed in the infinitesimal range that allows complex differentiation of matter to occur. Of course, observers are unlikely to exist in universes that contain only a single black hole or a vast sparse void. The idea that human existence necessitates that the Universe satisfy very particular conditions is referred to as the *anthropic principle*.

Within individual clumps of matter the same problem arises—negative total energy leads to collapse, positive energy means that the object expands, and the range of stable objects appears to be infinitesimally small. The existence of stable objects requires an equilibrium, in which outward forces balance gravity. This equilibrium must be stable, with chance outward motion resulting in a restoring inward force, and vice versa. These outward forces vary depending on the object—for small objects, up to planetary size, gravity is sufficiently weak that material strength and chemical bonds dominate and can stop any collapse. In stars, internal gas pressure is the countervailing force, and the stable equilibrium thus obtained is demonstrated by the existence of many billions of stars in each of many billions of galaxies. Given the success of our particular Universe in forming stable individual objects, the question is reversed: In a Universe that routinely generates stable objects in a dizzying array of shapes and sizes, under what circumstances can black holes form?

6.1 Stellar-Mass Black Holes

Stellar-mass black holes are generally understood to be created in supernova explosions that mark the end of the life of a massive star.[1] Many supernovae create neutron stars rather than black holes, and the precise conditions under which black holes form are still not fully understood. If the black hole is to be detected, further events are required, such as the formation of a binary star system of a kind that can be observed, and in which the existence of a black hole can be demonstrated. Here we discuss first the supernova event and then the binary star evolution that leads to a situation in which the black hole can be observed.

6.1.1 Black Holes from Supernovae

Stellar mass black holes are the final stage in the life cycle of massive stars. The key event that creates the black hole is a supernova explosion. Prior to the supernova, the star evolves by generating energy through fusion of light elements into heavier elements, starting with hydrogen into helium. In a low-mass star, this process terminates when the stellar core is dominated by carbon and oxygen (sometimes oxygen, magnesium, and neon) supported by Fermi degeneracy pressure. Once the outer envelope of the star is ejected, the core remains as a white dwarf. Significant amounts of mass can be lost by the star as it evolves, so a star that ends up as a white dwarf does not

---

[1] The term *supernova* is used to denote two kinds of events. A Type Ia supernova occurs when a white dwarf accumulates sufficient mass to exceed the Chandrasekhar limit. These supernovae have luminosities that can be precisely determined by their observable characteristics and are thus extremely useful as cosmological markers. They undergo explosive nucleosynthesis that blows the star apart and thus does not leave behind a compact remnant. Here we are referring to the "core-collapse" supernova that occurs at the end of the life of a massive star, which under most conditions does leave behind a compact remnant.

necessarily spend most of its life with a mass less than the Chandrasekhar limit. It is currently believed that stars with an initial mass of less than 8 $M_\odot$ evolve into white dwarfs, whereas stars more massive than 8 $M_\odot$ undergo a supernova explosion.

In contrast with low-mass stars, the cores of massive stars become hot enough for carbon to fuse into heavier elements, which leads to a succession of fusion processes culminating in a stellar core dominated by iron. Iron does not generate energy from fusion, so further energy production is not possible. The iron core can thus be supported only by Fermi degeneracy pressure. Outside the core, material from the star continues to be fused into iron, so the degenerate core grows quickly. When the iron core exceeds the Chandrasekhar limit, it collapses, setting off the chain of events that results in a supernova.

To collapse, the core must lose most of its electrons, which it does through the process of electron capture—electrons and protons combine to form neutrons and neutrinos. The result is a very small core of approximately the Chandrasekhar mass composed largely of neutrons. The neutrinos stream outward, carrying with them much of the energy associated with the gravitational collapse of the core.

The key to the formation of a black hole lies in the behavior of the material outside the iron core. It must initially collapse onto the core. This collapse generates gravitational energy, and this additional energy is transferred into the increasingly dense outer material. The neutrinos streaming outward from the newly formed neutron star also interact with the outer material, injecting energy into the plasma.

All this energy generates fusion processes in the outer material, leading rapidly to the formation of heavier elements, including elements heavier than iron. In many cases, the total amount of energy generated in or imparted to the regions of the star outside the iron core are sufficient to gravitationally unbind that material. Effectively, the outer regions of the star "bounce" off the newly formed neutron star and are ejected into the interstellar medium. This ejected material radiates strongly, largely from radioactive decay of the elements created in the supernova event itself, generating the intense emission observed from supernovae. The ejected material forms a supernova remnant, expanding and eventually merging into the interstellar medium. Thus most core-collapse supernovae result in a central neutron star and an expanding supernova remnant.

In some cases, however, the outer layers may not be ejected but may fall back onto the compact object at the center. It is this situation that gives rise to the creation of a black hole. Such fallback of the outer regions of the star is more likely when the progenitor star immediately prior to the supernova is more massive—the energy required to unbind all the outer regions of the star is greater for more massive stars. Thus in general terms, there is a progenitor mass below which a supernova gives rise to a neutron star and above which sufficient mass falls back to create a black hole. This mass boundary between supernovae that create neutron stars and those that create black holes is thought to be between 20 and 40 $M_\odot$.

The precise conditions under which a black hole, rather than a neutron star, is born is not currently well understood and is a subject of considerable current research.

The structure of the star immediately prior to collapse and, in particular, its chemical composition and rotation as a function of radius are critical to the outcome of the supernova. However, the evolution of massive stars is imperfectly understood, in part because the crucial role of mass loss from stellar winds is not well determined. The presence of a binary companion star can often also have a significant effect on the structure of the presupernova progenitor, particularly if mass is exchanged between the two binary components. Even if the progenitor structure were perfectly understood, however, the results of the supernova event would be hard to determine. Detailed computer models are used to predict the outcome of supernova events. Such computations must include the effects of energy transport by neutrinos and a wide variety of nuclear fusion and fission processes, all in the context of a process in which crucial events take place on scales as small as a few kilometers near the neutron star surface and as large as the outer regions of the expanding supernova remnant, which can start at the outer edge of the progenitor star, hundreds of millions of kilometers from the core, and expand from there. Recently there have been enormous improvements in computer hardware and significant progress in the sophistication of the programs used to model supernova events. Nevertheless, details of the relationship between the initial mass and structure of a massive star, its mass and structure immediately prior to the supernova event, and the consequences of the supernova in terms of the expanding supernova remnant and the nature and the precise mass of the compact remnant left behind are still poorly understood.

6.1.2 Binary Star Evolution

To be observed as an X-ray binary, not only must a black hole be created in a supernova event but subsequent evolution must result in a companion star providing a stream of matter to the black hole so that significant numbers of X-rays can be observed. The details of binary star evolution must therefore also be considered in determining the number and kind of observable stellar-mass black holes.

For high-mass X-ray binaries, the situation is relatively straightforward. These are systems in which a black hole is accreting from a star more massive than itself that emits a strong stellar wind. This kind of situation evolves naturally from a binary consisting of two massive stars. The more massive star (the *primary star*) evolves faster and thus undergoes a supernova explosion first. Some fraction of the mass of the primary is lost to the system, while the rest collapses into a compact object. For circular orbits, $V^2 = GM/D$, so the kinetic energy (proportional to $V^2/2$) is half that of the gravitational binding energy (proportional to GM/D). Thus if half of the mass is lost to the system, the binding energy and the kinetic energy are the same; if more than half of the mass is lost, the kinetic energy is greater than the binding energy, the total energy is positive, and the system becomes unbound. Even if the system remains bound, the mass lost from the supernova can dramatically change the parameters of the binary system.

However, there is another crucial phenomenon that must be understood to determine the parameters of the binary that emerges from the supernova, namely, the mass loss and the neutrino flow from the supernova is likely to be asymmetric, which results in a "kick" being imparted to the nascent black hole as more material is ejected from one side of the supernova than from the other. The observed distributions and velocities of known neutron stars and black holes demonstrate that such kicks do indeed exist and can be as much as hundreds of kilometers per second. If the kick is in the right direction, a binary that would otherwise be unbound can remain intact, while a binary that would otherwise survive can be disrupted. But in any case, the kick will certainly alter the orbital period and eccentricity of the binary.

Provided the binary remains intact after the supernova event, with an orbital period of less than a year or so, then the high stellar winds emitted by the secondary star will essentially guarantee the creation of an observable X-ray binary. Such a high-mass X-ray binary will be relatively short-lived (only a few million years), since the secondary star will also evolve quickly and will undergo a supernova of its own. Thus they will not have time to move far from their initial birthplace, and indeed high-mass X-ray binaries in our own and nearby galaxies are observed to be in or very near regions of current high-mass star formation. If the kick is in an appropriate direction, this second supernova can result in the formation of a binary system containing two compact objects. Double neutron star binaries in which one object is a radio pulsar are observed and will be discussed further in chapter 9 in the context of the gravitational wave radiation they emit. Binaries consisting of a black hole and a neutron star, or of two black holes, have not yet been definitively observed but are expected to exist in large numbers.

The evolution of low-mass X-ray binaries (those in which the companion star is less massive than the compact object and transfers mass through Roche lobe overflow) is more complicated than that of high-mass systems. Since the companion star is relatively low mass, it is almost inevitable that the system will lose more than half its total mass in the initial supernova explosion. Thus the binary will remain bound only if the kick has a fairly precise size and direction. Because low-mass stars emit very modest stellar winds, if the binary does remain intact, the object will have quite low X-ray emission, and thus be difficult to observe, unless the binary orbit is short enough that the companion star fills its Roche lobe—this will require an orbital period of less than a day for typical main sequence stars.

A number of effects tend to extract energy from the binary orbit and thus reduce the period. Of these, *magnetic braking* may well be the strongest effect. Magnetic braking occurs when material emitted from a rotating star is forced by the star's magnetic field to corotate with the star. This is an example of a magnetic-dominated fluid flow in which the gas is required to flow along the magnetic field lines. As the material moves outward its angular momentum must increase, because the angular velocity remains the same, but the distance from the central source increases. This angular momentum is extracted from the star, slowing its rotation. In a close binary system, tidal effects force the rotation of the star into synchronicity with the orbit, so the loss of angular momentum is coupled into the orbit, thus driving the stars closer together and decreasing the orbital period.

Alternatively, the companion star may come into contact with the compact object when it evolves into a giant, the star which can fill the Roche lobe of orbits with periods of up to several hundred days. In this case the angular momentum of the orbit is maintained, so the mass transfer is conservative, and the binary gradually widens as mass is transferred from the lower-mass giant to the higher-mass compact object. However the continuing evolution of the giant causes it to increase in size as well, so the mass transfer is maintained. X-ray binaries are observed with both main sequence and giant companions, so both of these effects must be relevant to the creation of low-mass black hole binaries.

Conservative mass transfer requires that mass transfer from a low-mass to a high-mass object should increase the orbital period, If the mass transfer is to continue, the system must either continue to lose angular momentum (for main sequence companions), or the star must continue to increase in size (for giant companions). The rate of mass transfer is determined by the rate of the angular momentum loss or the size increase of the giant. These processes have long timescales, up to billions of years. Thus low-mass X-ray binaries are much longer-lived than their high-mass counterparts and are often found in old stellar populations with very little current star formation. Interestingly, in galaxies like the Milky Way, there appear to be roughly equal numbers of high-mass and low-mass systems, implying that the longer lifetimes of the low-mass systems are compensated for by the more stringent conditions of their formation.

6.1.3 Binary Evolution in Star Clusters

A particular situation arises in dense star clusters. Here the evolution of binary stars can be dramatically changed by close encounters between the closely packed cluster stars. If a single star passes close to or within the orbit of a binary star, several outcomes can lead to systems that could not occur under ordinary circumstances. The results of encounters between binaries and single stars depend strongly on the total energy of the three-body system including the binary and the incoming single star. If the negative energy of the binary exceeds the kinetic energy of the single star, then the total energy will be negative, and the binary is referred to as a *hard binary*, whereas if the total energy of the three-body system is positive, the binary is described as *soft*. Possible outcomes of close encounters between binary stars and single stars include the following:

- The orbit of the binary star can be altered. This alteration can be minor or quite drastic, depending on the details of the encounter. In general, encounters between single stars and hard binaries tend to cause the binary to become more tightly bound—that is, hard binaries tend to become harder. In this case, the single star leaves with more energy than it had originally, providing a source of energy to the dynamics of the cluster as a whole. This is true only in a statistical sense, that is, the average effect of such encounters is to make hard binaries harder, but any particular encounter can change the binary parameters in many different ways. But the net effect in globular clusters is to bring tight binaries closer together, which can lead to the initiation of mass transfer sooner than would otherwise be the case. In contrast, loosely bound binaries, in which the relative velocity of the binary components is smaller than the velocity of the incoming star, tend to end up more loosely bound than before the encounter.

- The binary can be completely disrupted; that is, the three stars can end up independent of one another. This can occur only if the total energy of the system is positive, and thus this outcome requires that the binary be soft. Since soft binaries become generally softer and eventually become disrupted, and hard binaries become harder, this tends to create a situation in which all binaries in a cluster are tightly bound, with short orbital periods, in contrast with binaries in the field (outside of clusters), which show a wide range of orbital periods.

- The single star can be exchanged into the binary. In such an *exchange encounter*, the incoming single star ends up as part of the binary system, and one member of the binary is ejected. In general, exchanges tend to result in retention of the more massive star of the binary system and ejection of a lower-mass star. In systems where binaries consist of low-mass main sequence or giant stars and a large number of massive compact remnants, exchange encounters tend to result in the creation of large numbers of binaries containing massive compact objects. This effect is likely to be important in explaining the disproportionate number of such binaries in dense star clusters.

Binary stars can also encounter other binary stars, in which case the range of outcomes is very broad. But the general trends are the same as for encounters between single stars and binaries: hard binaries harden; soft binaries soften and are often disrupted, and tight binaries containing the most massive stars tend to be produced.

All these effects are amplified by the process of *mass segregation*. Once the cluster has gone through at least one relaxation time, the massive stars in a cluster tend to sink to the middle of the cluster, while lower-mass stars migrate toward the outer regions of the cluster and are sometimes expelled from the cluster altogether. Since hard binaries segregate in the same way as single stars with mass equal to the total of the two components of the binary, mass segregation tends to create a dense core in the cluster consisting of binaries and the most massive stars, which are often compact objects. Since the cluster core is dense, there are frequent interactions between the binaries and the compact objects, further increasing the tendency for binaries including compact objects to form.

These effects explain why dense star clusters contain many more X-ray binaries per unit mass than the general galactic field. It is curious, however, that all the luminous

X-ray binaries known in galactic globular clusters exhibit X-ray bursts and thus must contain neutron stars rather than black holes. This observation has generated some discussion of the fate of black holes in a cluster in which most stars are below 1 $M_\odot$, as would be the case in an old star cluster. It is clear that the massive remnants would undergo significant mass segregation, effectively forming a cluster of their own at the very center of the cluster. How such a subcluster of massive objects evolves is the subject of much current debate—one possibility is that they tend to kick each other out of the cluster, leaving perhaps only a single object or a single massive binary behind.

6.2 Supermassive Black Holes

In contrast with stellar-mass black hole formation, there is no obvious route to creation of a black hole of $\geq 10^6 \, M_\odot$ directly from collapsing interstellar gas. It seems unlikely that a single object of such a mass could form without fragmenting into individual stars. Thus most discussions of the origin and evolution of supermassive black holes posit an initial "seed" black hole of relatively low mass, which then grows over time. While this general scenario seems plausible, the details are not yet clear. In particular, because supermassive black holes are observed at high redshifts, and thus at relatively early times in the evolution of the Universe, this process must progress surprisingly quickly in at least some cases.

6.2.1 Seed Black Holes

It is generally assumed that the seed black holes that subsequently evolved into supermassive black holes must have originated as stellar-mass black holes in the earliest generation of star formation. This assumption presents some difficulties, as the growth mechanisms postulated next cannot create supermassive black holes out of $\approx 10\ M_\odot$ black holes quickly enough to explain the AGN observed at high redshifts. However, the very first generation of stars is likely to have been very different from the stars observed today. In particular, the first generation of stars was presumably created out of a mixture of hydrogen and helium, with no heavier elements at all, since the elements heavier than helium could not be created in the early universe but, rather, required nuclear processes inside stars to come into being and supernova explosions to be distributed into the interstellar medium.

Stars without heavy elements are not currently observed, so this first generation of stars has apparently disappeared. However, theoretical models of the behavior of such stars suggest that the absence of even trace amounts of heavy elements changes the nature of the stars quite dramatically. In particular, much of the opacity and cooling in stars and star-forming regions is generated by incompletely ionized heavy elements. In the absence of such ions, the mass of the fragments into which a collapsing gas cloud forms is larger than it might otherwise be. Stellar winds are also harder to drive, so more of the mass is retained as the star evolves. Finally, the eventual supernova of such an object at the completion of nuclear burning might plausibly lead to near-complete collapse, leaving behind a black hole much more massive than ones created by ordinary stars, perhaps in some cases with a mass as high as $10^3 \, M_\odot$. This scenario also has the appealing feature that the large mass of the first generation of stars explains why no such stars are currently observed, since high-mass stars evolve and die on timescales much shorter than the age of the Universe.

6.2.2 Growth of Supermassive Black Holes

As discussed in the preceding section, it is plausible that there might be a significant number of black holes with masses of up to $10^3 \, M_\odot$ that were created very shortly after the first generation of stars was formed. However, black holes of this mass are not observed (see chapter 8) but, rather, must grow by factors of thousands to create the supermassive black holes observed back to very early times in AGN. While the details of the growth of supermassive black holes are still the subject of intense research and debate, two basic processes are generally assumed to be responsible.

The first growth mechanism is simply accretion. It is known that these black holes are accreting gas, since it is through the accretion process that they are observed. If a steady supply of gas can be provided to the black hole, its mass will inevitably grow. However, accretion cannot in general occur faster than the Eddington limit will allow— if mass is accreted too quickly, the luminosity will exceed $L_{\rm Edd}$, and the accretion will be halted by radiation pressure.

Since $L_{\text{Edd}} \propto M$, the allowed mass accretion rate is a constant proportion of the mass, leading to exponential growth with an e-folding timescale of

$$\tau_{\text{growth}} = \eta \frac{\sigma_T c}{4\pi \, G m_p} \approx 5 \times 10^7 \text{years},$$

where η is an efficiency factor for turning accreted mass into radiation, usually taken to be about 10%. It might be thought that exponential growth could account for even very large black hole masses, but in fact the number of available e-folding times is quite limited. Bright quasars have been observed at redshifts greater than 6, which implies distances so great that the Universe was $<10^9$ years old when the quasars were observed. This is also about the time when the first stars were formed, which produced the seed black holes at the end of their evolution. Thus there may be 10 or fewer e-folding times between the creation of the seeds and the existence of bright quasars that seem to require $10^9 \, M_\odot$ SMBHs. Since $e^{10} \approx 2 \times 10^4$, it may be hard to grow a black hole from $10 \, M_\odot$, or even 10^3–$10^9 \, M_\odot$, with Eddington-limited accretion.

The source of the gas that powers AGN is also not wholly clear. There is considerable interstellar gas in most young galaxies, but it is not easy to extract enough angular momentum to allow a substantial amount of the gas to fall into the small event horizon of a black hole. Torques exerted by nonspherical mass distributions in the center of galaxies (referred to as "bars") may help this process, but it is not clear that it can provide enough gas to grow the black holes quickly enough. Another source of gas could be tidally disrupted stars. When a star ventures too close to a massive black hole, it is torn apart by tidal forces and stretched out into a stream of gas. Such a stream is accreted onto the black hole relatively quickly. However, as with the direct accretion of gas, ensuring a steady supply of stars to the black hole can be problematic.

The other basic idea relating to black hole growth is merger of black holes. This concept has gained currency since the discovery of the M-sigma relationship, which demonstrates that the mass of the central black hole is proportional to the mass of the spherical component of the galaxy that contains it. Galaxies are thought to grow by mergers as small condensations of gas and stars collide and create increasingly larger structures (see figure 6.1). Thus it seems natural to ask whether the black holes they contain might also merge. If the galaxies and black holes grow together, then the close relationship between them could be explained.

However, the processes through which the central black holes in merging galaxies will themselves merge are also not well understood. In general, the galaxies will not collide precisely head-on, so the two central black holes will start out in a fairly wide orbit around each other. They will gradually spiral inward, by transferring energy from their orbit to the stars and gas they encounter— this is essentially an extreme example of the process of mass segregation that occurs in star clusters. However, this process depletes the inner regions of the galaxy of matter and is thus self-limiting. If the two black holes are brought sufficiently close to each other, they will merge through the emission of gravitational radiation, an event that could

Figure 6.1. Image of colliding galaxy SDSS J1254+0846 in optical and X-ray light. The outer parts of the galaxy, observed in optical light, are tidally distorted. Each galaxy contains an AGN, observed in X-rays, which are the bright sources in the center. Both the galaxy and the black holes they contain are likely to merge over the next few hundred million years. X-ray image NASA/CXC/SAO. P. Green et al., Optical image (Carnegie Obs/Magellan/W. Baade Telescope. J.S. Mulchaey et al.).

potentially be observed by gravitational wave detectors currently under development. Unfortunately, according to simple calculations, the interaction of orbiting black holes with stars and gas in typical galaxies does not appear to bring them close enough for gravitational radiation to take them the rest of the way to merger. This "final-parsec problem" is a current subject of considerable research.

One interesting test of whether gas accretion or black hole mergers is the dominant process would be to compare the total luminosity of AGN in the Universe with the total mass of AGN.[2] If most of the mass of the SMBHs is due to accretion, then the mass accretion responsible for the observed luminosity must be equal to the current total mass of the AGN. However, if there is more mass than can be accounted for by the observed accretion luminosity, then some "quiet" mode of growth might be required. Black hole mergers constitute such a quiet growth process, since they emit much of their energy in the form of gravitational waves, which are currently not detected (but see chapter 9). There are significant observational and theoretical problems with carrying out this test. Observationally, one must include all energy emitted by AGN, including radiation at all wavelengths and bulk energy carried out in jets. Theoretically, one must understand the efficiency of the accretion, that is, how much energy is generated by the accretion of a given amount of mass. These complexities are not yet wholly resolved.

Thus our understanding of the creation of supermassive black holes is currently incomplete. Observational approaches to the problem include determining the luminosity of AGN as a function of cosmic time by detailed surveys at all redshifts; searching for the first quasars and stars at the highest redshifts to try to identify black holes

[2]This approach was pioneered by A. Soltan (1982) in an article in *Monthly Notices of the Royal Astronomical Society* 200:115.

in the initial or early stages of their growth; searching for binary black holes in merging galaxies; and creating gravitational wave detectors that might one day detect the merger events themselves. Theoretical work is being done on the formation and evolution of the earliest generation of stars; on the processes by which gas might be fed to growing black holes in the centers of galaxies; and on processes that might bring merging black holes across the final parsec of their separation so that mergers can take place.

7

DO INTERMEDIATE-MASS BLACK HOLES EXIST?

There is powerful empirical evidence for two classes of black holes, namely, the stellar-mass black holes, with masses a few times that of the Sun, and the supermassive black holes at the centers of galaxies. The considerable gap in mass between these two categories naturally prompts the question whether black holes might exist at other mass scales. In recent years two lines of evidence have been presented in support of the idea that black holes with masses intermediate between stellar mass and supermassive might exist, that is, with masses of 10^2–10^5 $M_\odot$. Such sources are referred to as *intermediate-mass black holes* (IMBHs). In both cases the results are currently still ambiguous, and much debated.

7.1 Ultraluminous X-Ray Binaries

As X-ray astronomy evolved, one of the capabilities that improved significantly was the *spatial resolution* of the various orbiting observatories, that is, the ability of an observation to distinguish between closely neighboring objects. Improved X-ray optics and detector technology has improved spatial resolution of X-ray observatories to the point that the spatial resolution in soft X-rays of the *Chandra* satellite is almost as good as that of the Hubble Space Telescope in optical bandwidths.

One scientific goal for which the improved spatial resolution has been particularly important is the study of X-ray binaries in galaxies other than our own. X-ray binaries are bright, so simply detecting them in other galaxies in the nearby universe is straightforward. However there are many such objects in a sizable galaxy (we have hundreds in our own), and they are generally concentrated toward the central regions, or in the regions of the greatest ongoing star formation. Those parts of galaxies are also home to other strong X-ray sources, including AGN and supernovae remnants. Thus studying the populations of X-ray binaries in other galaxies requires distinguishing many closely packed sources from each other and thus requires the highest obtainable spatial resolution.

Maps of nearby galaxies made with *Chandra* and other X-ray observatories have provided lists of hundreds of X-ray binaries outside the Milky Way. The demographics of the X-ray binary population vary from one galaxy to another. One particular category of X-ray binary that has been identified in other galaxies, but not in our own, is the *ultraluminous X-ray sources* (ULXs), whose luminosities are significantly larger than the Eddington limit for a 10 $M_\odot$ object (see figure 7.1).

When the first ULXs were observed, there was some question about whether the large observed fluxes emanated from an individual X-ray binary. And, indeed, as

Figure 7.1. X-ray image of galaxy M66. The point sources are X-ray binaries superposed on a diffuse galactic background. The brightest sources are considerably brighter than the Eddington limit of a 10 $M_\odot$ accreting object and so are considered ultraluminous sources. NASA/CXC/Ohio State Univ. C. Grier et al.

observations were refined, some sources were identified as AGN, or as superpositions of several individual sources, or as foreground or background sources coincidentally located within the galaxy under study. But some ULXs could not be explained in this way. Such sources are not coincident with the center of the galaxy (as an AGN would be) and have spectra that are hard to explain from foreground or background sources. A key step forward was the identification of large luminosity changes over time from some of these sources—if the flux came from several superposed sources, luminosity changes of more than a factor of 2, would not be expected since that would require the individual sources to be able to arrange to vary in concert. Thus at least some of the ULXs do appear to be individual sources that have luminosities of $> 10^{40}$ ergs^{-1}, well above the Eddington limit for stellar-mass black holes.

The natural suggestion for these ULXs is that they represent accretion onto objects significantly more massive than the accreting black holes observed in our galaxy. Since the Eddington limit scales with mass, the ULXs could remain within the Eddington limit if the accreting sources contained intermediate-mass rather than stellar-mass black holes. This hypothesis was strengthened by the discovery that the X-ray flux from ULXs is dominated by soft (low-energy) photons. Radiation that is thermal in origin would indicate lower temperatures from the accretion disk, in accordance with the expectation that $T \propto M^{-1/4}$ for accreting black holes.

However, detailed studies of the X-ray spectra of the ULXs do not agree with the expectations of the IMBH hypothesis. If the ULXs were intermediate-mass black holes, they would presumably be accreting at sub-Eddington rates, and thus the X-ray spectra should show states similar to those of the known galactic X-ray binaries, except scaled appropriately for the difference in mass of the black hole. However, at least some of the ULXs have spectral features that are different from anything else seen in the galaxy. In particular, some of the best-observed ULXs show an excess of flux in low-energy X-rays, even when the $T \propto M^{-1/4}$ law is accounted for, and also a sharp drop in flux at higher energies unlike anything else seen in galactic sources.[1]

---

[1] J. Gladstone, T. Roberts, and C. Done, 2009, *Monthly Notices of the Royal Astronomical Society* 39:1836.

This suggests that the accretion flow in ULXs is not analogous to what is observed in the galactic X-ray binaries.

The alternative hypothesis is that the ULXs are examples of super-Eddington accretion onto stellar-mass black holes. Super-Eddington accretion requires that one or more of the assumptions that underlie the calculation of the Eddington limit be false. In particular, the flow is unlikely to be spherically symmetric, so from some viewing angles luminosity might be seen well beyond the nominal Eddington limit. The flow is also likely to become optically thick, which might result in emergence of observed radiation at lower energies, which could explain the observed spectral anomalies. Thus super-Eddington accretion onto a stellar-mass black hole may be a viable explanation for the ULXs.

A compelling resolution to the ULX conundrum is likely to require direct determination of the mass of the accreting object. This could in principle be done by determining the orbital period and velocity of the companion star. In practice, this is quite difficult: the sources are located in external galaxies, and thus the companion stars are extremely faint. At the same time the accretion luminosity is by definition large, so the optical/infrared flux from the companion star is likely to be masked by that of the accretion. However, such measurements may not be out of the question if the companion is a bright early-type giant or supergiant, as has been suggested for some sources. If such a source goes into a low state in which the companion outshines the accretion flow, the orbital motion of the companion might be measured, and the existence of an intermediate-mass black hole could be established or refuted.

7.2 Black Holes in Star Clusters and Low-Mass Galaxies

The putative intermediate-mass black holes in ULXs are fueled by a companion star and are thus similar to X-ray binaries. One can also imagine intermediate-mass black holes at the center of low-mass galaxies and star clusters, which would thus be low-mass analogs to the supermassive black holes in AGN. The mass distribution of black holes at the center of galaxies is likely to extend well below $10^6\ M_\odot$. Observations in nearby galaxies seem to support the presence of black holes at the low end of the AGN mass distribution, although there is some evidence that the M–σ relationship changes at the low-mass end. But the relevant observations are hard to make in very low-mass galaxies, so the low-mass extension of the SMBHs has been hard to characterize.

If the M–σ relation extends to globular star clusters, the largest of which contain $10^{6-7}\ M_\odot$ in stars, then these systems might be expected to contain central black holes that are intermediate between the stellar-mass and supermassive black holes. Such black holes would be distinct from the compact objects contained in X-ray binaries in these star clusters, as they would presumably not have evolved as members of a binary star system and would be located at the exact dynamic center of the cluster. Since there is no evidence of X-ray sources with $L_x \gg 10^{38}\ \mathrm{ergs}^{-1}$ in these systems, such central IMBHs apparently do not accrete significant amounts of matter. This is not surprising, since globular clusters are known to contain only very small amounts of interstellar gas. It is possible that newly developed sensitive radio telescopes may be able to detect such accretion, and, indeed, one cluster in the nearby Andromeda galaxy seems to show unexpectedly large radio flux. Barring direct evidence of accretion, demonstrating the existence of IMBHs would require seeing the gravitational effects of the IMBH on the dynamics of observable stars near the center of the cluster.

Some evidence of this kind has been claimed, in the form of *cusps* of unexpectedly high density and velocity near the center of some clusters. These cusps are analogous to the density peaks at the center of galaxies that reveal the presence of nonaccreting central black holes. The outer edge of the cusp occurs when the mass of stars interior to that point is roughly equal to the mass of the IMBH. Thus the total mass of stars exhibiting a significant influence from the IMBH cannot be greater than the mass of the IMBH itself. An IMBH with $M \approx 10^3 \, M_\odot$, for example, would influence no more than the central 10^3 stars in the cluster if the typical mass of a star is close to that of the Sun. Since some of these stars are quite faint, there may be only a handful of easily observable stars that would show the effects of the IMBH. These effects may be quite subtle, and there is disagreement in the scientific community as to whether the claimed evidence for IMBHs in globular clusters might be explained by random fluctuations in the distributions of the small number of stars, expected to be influenced by an IMBH. For example, in the largest cluster in our galaxy, Omega Centauri, ground-based observations seem to show a cusp of the kind expected to be produced by an IMBH, whereas space-based observations do not. One possible resolution to this and other discrepancies lies in ambiguities in determining the location of the center of the cluster; if the cluster center is determined by the distribution of light, it tends to be unduly influenced by the location of the brightest few stars near the center. If the measured center of light is superposed on such a bright star, the light distribution will naturally have a cusp, regardless of whether there is a central IMBH or not.

To firmly establish (or refute) the presence of IMBHs in globular clusters will require very precise measurements of as many stars as possible. Two approaches may be helpful: first, repeated images by the Hubble Space Telescope can reveal small motions across the sky by many stars in the central region of clusters. But even in this case, the limits on the numbers of stars influenced by the IMBHs may preclude conclusive results. Alternatively, the motion of radio pulsars may be very precisely determined by changes in the observed pulsation periods—if such a pulsar is close enough to an IMBH, the changes in the motion of the pulsar may reveal the presence of a large compact mass. This approach would require that a pulsar exist in just the right position near the central black hole and is thus also unlikely to provide definitive results. Thus it seems likely that the existence of IMBHs in star clusters will remain unproven in the foreseeable future.

8

BLACK HOLE SPIN

Black holes are among the simplest objects in the Universe. Simplicity and complexity can be defined by the number of parameters required to completely specify the properties of an object. This number is very large for such things as people, planets, stars, and galaxies—the chemical composition, pressure, and temperature all need to be defined at every point within the object. But a black hole can be completely defined by a mere three parameters: its overall mass, charge, and spin. The nature and distribution of material inside the event horizon has no impact on the observable properties of the black hole.

The previous chapters discussed ways in which the mass of black holes is determined and the evolutionary scenarios that lead to black holes with masses similar to those observed. Charged black holes are not likely to occur in astrophysical situations, since a black hole with nonzero charge would attract oppositely charged particles and would quickly achieve neutrality. But observed black holes are expected to have nonzero angular momentum. In particular, since black holes are created by the collapse of large amounts of material, any initial spin will be increased during the collapse due to angular momentum conservation. So the spin of a black hole is likely to be significant.

As noted in chapter 1, the spin is usually described as a nondimensional parameter $a = J/(GM^2/c)$, which can range from zero (a nonspinning black hole) to 1 (a situation described as "maximally spinning"). The parameter a enters into the Kerr metric, which describes a black hole with a ring singularity at the center, surrounded by an event horizon, surrounded by an *ergosphere*. Between the event horizon and the ergosphere, objects cannot remain at rest but must rotate along with the black hole; nevertheless, they can escape to infinity. If pairs of particles are produced in the ergosphere, with one particle plunging into the event horizon and the other escaping to infinity, the rotational energy of the black hole can be extracted. This mechanism is known as the *Penrose process*. Up to 29% of the total energy of a maximally rotating black hole can be extracted by this process if the entire spin of the black hole is removed.

The differences in space-time between a nonspinning Schwarzschild black hole and a Kerr black hole of the same mass have potentially observable effects. Thus we can hope to measure the spin by observing effects which should be different depending on the value of a. The most obvious of these differences is the position of the *innermost stable circular orbit* (ISCO), which has a significant effect on the inner edge of an accretion disk. It is through determination of the physical size of the ISCO that the spins of black holes are determined.

8.1 The Innermost Stable Circular Orbit

Accretion disks are composed of gas orbiting in concentric circular orbits around the accreting object. Viscosity extracts energy from the disk and transports angular momentum through it, which has the effect of transferring mass toward the inside of the disk. This process continues until the accreting matter reaches the inner edge of the disk. Three things can truncate the inner disk. The disk can reach all the way down to the surface of the accreting object, at which point the accreting gas must settle on that surface. The disk can also terminate due to the presence of strong magnetic fields anchored in the surface of the accreting object. When the magnetic pressure becomes sufficiently great, the gas must flow along the magnetic field lines directly to the surface of the accreting object. These fields change the balance of forces and thus disrupt the orbits of the gas. But black holes have no surface, and so neither of these effects will terminate the disk.

Instead, in the case of accretion black holes the properties of space-time itself bound the inner edge of the disk. Sufficiently close to a black hole, stable circular orbits are no longer possible. For a nonspinning black hole described by the Schwarzschild metric, this innermost stable circular orbit, or ISCO, occurs at $r = 3R_s$. Between the ISCO and the event horizon, material can follow trajectories that do not lead inside the event horizon, but those trajectories cannot remain at a fixed distance from the black hole, so circular orbits are impossible.

The same is true of rotating black holes, described by the Kerr metric. However, the distance of the ISCO from

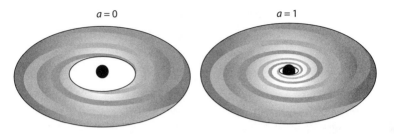

Figure 8.1. Illustration of the innermost stable circular disk (ISCO) for a nonspinning and maximally spinning prograde black hole. The accretion disk terminates at the ISCO.

the singularity varies with the spin of the black hole. As the spin parameter a goes from 0 to 1 the distance of the ISCO from the center of the black hole decreases from $6GM/c^2$ to $1GM/c^2$. Thus the accretion disk extends farther in for a spinning black hole than for a nonspinning black hole as shown in figure 8.1. This result applies to situations in which the disk is rotating in the same direction as the black hole. Relativistic effects force the inner parts of the disk to orbit in a plane perpendicular to the angular momentum vector of the black hole. However, the disk can be prograde or retrograde: that is, the disk can be rotating in the same direction as the black hole or in the opposite direction, respectively. The prograde situation results in a small ISCO, whereas retrograde rotation creates a much larger ISCO. Either way, the temperature of the inner disk is quite different for a disk with $a \approx 1$ than for a nonspinning disk. This difference can be observed because the inner edge of the disk is the hottest part of the disk, so the location of the high-temperature cutoff in the spectrum will change if the ISCO is in a different position. Another potentially observable effect is that light from the innermost parts of an accretion disk that extends closer to the black hole will be more gravitationally redshifted.

There is some debate over the correct way to model the effects of the ISCO. The simplest approach is simply to assume that the material vanishes down the black hole as soon as it reaches the ISCO. But this is obviously not quite right—the density of the material doubtless drops considerably at the ISCO, but the material still has to make its way to the event horizon before disappearing from sight, so the density and emission must in fact be continuous across the ISCO. It may be that the drop is quite precipitous, but some dynamical models suggest that material inside the ISCO still contributes significantly to the observed luminosity of the disk. The best way to model the flow from the ISCO to the event horizon is a subject of considerable dispute, leading to strong disagreements on the reliability of measurements of black hole spin.

8.2 Observations of the ISCO through Line Emission

The spin of a black hole can be determined by measuring the size of the ISCO relative to R_s. There are two primary measurement methods, both of which have been carried out in some cases. But there is considerable controversy over whether either of these methods is reliable, a controversy that has been heightened by the quite different answers obtained in the very few cases in which both methods have been applied.

The first approach to measuring black hole spin involves observing spectroscopic emission lines from the inner parts of the accretion disk. Spectral lines are generated by atomic or nuclear processes and consist of spikes of emission at a specific wavelength or, equivalently, at a specific energy or frequency. Accretion disks sometimes display such lines. Highly ionized iron, in particular, has a number of spectral features in the X-ray range. Since iron is the most massive nucleus found in abundance in astronomical plasmas, these lines are the highest-energy atomic lines generally observed and have been seen in accreting stellar-mass and supermassive black holes. A particularly useful and important line is a fluorescence line at an energy near 6.4 keV, which generally thought to be excited in the disk by photons emitted by the accretion disk corona. Thus this line is thought to track the gas in the disk.

Although the photons associated with a particular line are emitted at the same energy, they are not all observed at that energy. This phenomenon is called *line broadening*; that is the emission created by a spectral line is spread over a range of energies. If the spectrum of the source is plotted, the bump associated with the line has a particular shape determined by the line broadening (see figure 8.2). The physical mechanisms that give rise to the broadening can be determined by measuring the shape of the line.

In the case of line emission from an accretion disk, the Doppler shift results in considerable line broadening. Different parts of the disk move at different speeds relative to the observer, so the Doppler shift changes the observed energy of the photons differently at different points.

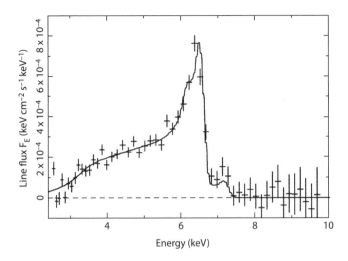

Figure 8.2. Iron line from the AGN MCG-6-30-15 obtained with the XMM satellite. Note the extended red wing, caused by gravitational redshift of the inner disk. The extent of the line is limited by the innermost edge of the accretion disk and can thus be used to determine the position of the ISCO. Source: Fabian et al. "A Long Hard Look at MCG-6-30-15 with XXM-Newton." MNRAS, 2002, volume 335, L1-L5. By permission of Oxford University Press.

When the disk as a whole is observed, the line is broadened because the velocity of the emitting gas varies across the disk. The internal motions of the disk generate a particular kind of line broadening, resulting in double-peaked emission—one peak corresponds to emission from the parts of the disk moving toward the observer, and the other, from those parts of the disk moving away from the observer.

But disks in strong gravitational fields display another effect, namely, broadening from gravitational redshift. The gravitational redshift changes the energy of photons due to the strength of the gravitational field from which a photon emerges. Like any other entity, a photon will lose energy as it climbs out of a gravitational potential well. In massive objects, this causes the object to slow down as it recedes from a large mass. But photons must by definition move at the speed of light, so they lose energy by changing wavelength rather than by changing speed. When the gravitating field is strong, as in an accretion disk near a black hole, the relativistic expression for the gravitational redshift for a Schwarzschild black hole as measured by an observer at infinite distance from the source can be expressed as

$$z = 1/\sqrt{1 - R_s/r} - 1,$$

where z is the observed redshift, and r is the distance from the massive central object to the source of emission.

The combination of Doppler shift and gravitational redshift produces a very particular line shape for line emission from an accretion disk near a black hole. The gravitational redshift moves the line to lower energies, but this effect is more pronounced for photons emerging from the inner parts of the accretion disk, which are nearest the black hole. This is also the part of the accretion disk in which the gas moves most rapidly, so the wings of the line profile are the most affected by the gravitational redshift. The high-energy wing moves back toward the center of the line, since the positive shift from the Doppler effect is countered by the gravitational redshift. But the low-energy wing becomes even more pronounced, as the Doppler shift and the gravitational redshift work in tandem. Thus the effect of the gravitational redshift is to make the line shape strongly asymmetric, stretched toward the low-energy end.

The amount of this stretching is determined by the strength of the gravitational field, that is, from the radial position of the emitting material relative to R_s. This is where the connection between line shape and the location of the ISCO, and thus the spin of the black hole, is made. For nonspinning black holes, the ISCO is relatively large, and thus the effects of gravitational redshift can never become too big. In contrast, the disk of a maximally spinning black hole extends much farther in, and thus the effects of gravitational redshift are much greater. Thus a rapidly spinning black hole will have lines with much greater asymmetry than a nonspinning black hole, so the spin of the black hole can be determined by measuring the shape of the lines.

There are a number of difficulties with this procedure in practice. Most important, the amount of line emission generated by a gas is determined by the physical conditions of the gas—in particular, the chemical composition, temperature, and pressure. The temperature and pressure vary greatly as a function of radius in an accretion disk, and thus the amount of line emission generated at each position in the accretion disk varies. One could imagine a disk in which the inner parts simply do not generate line emission, for example, if it is hot enough that the iron is completely ionized, and single-electron iron atoms are very few. The line shapes created by such a disk might mimic those of a disk with a much larger ISCO. Another problem is that accretion flows generate large amounts of line emission only in certain X-ray states, and those X-ray states may not be those in which the flow is in the form of a standard accretion disk. Thus the Doppler broadening may not be as expected from a disk. These effects can in principle be modeled, but the models are complex and not fully developed, so they introduce considerable uncertainty into the spin measurements. Nevertheless, studies of emission line shapes have produced spin measurements for a number of stellar and supermassive black holes.

8.3 Observations of the ISCO through Thermal Emission

A different approach to determining the location of the ISCO is to observe the continuous thermal emission from the hot gas in the inner parts of the disk. Specifically, a standard α-disk is expected to emit radiation that approximates blackbody radiation from a series of concentric rings, each of which has a different temperature. The temperature rises toward the inner edge of the disk by an amount that can readily be calculated. At the ISCO, the rise in temperature is abruptly terminated, so the highest temperature seen in the integrated spectrum of the disk corresponds to the temperature at the ISCO.

Blackbody radiation follows the Stefan-Boltzmann relationship $L = \sigma R^2 T^4$, which relates the luminosity, temperature, and size of the emitting region. The highest temperature of the disk can be determined by the high-energy cutoff of the observed spectrum. So, if the luminosity can be determined, the size of the emitting region can be calculated. For the highest temperature, that size will presumably be that of a ring of emission situated at the inner edge of the disk. The perimeter of that ring reveals the size of the ISCO. If the mass of the black hole, and thus R_s, can be separately determined, then the size of the ISCO in terms of R_s can be measured, and thus the spin of the black hole can be calculated.

Determining the luminosity of the disk emission requires knowing the distance to the black hole, so that the observed flux can be converted into an intrinsic luminosity. For supermassive black holes, the distance can be determined by measuring the redshift of the AGN or its host galaxy. But supermassive black holes present other difficulties to using this approach, as discussed later. Determining the distance to stellar-mass black holes is more difficult but can be done in a number of important cases.

Two approaches have been used to determine accurate distances to stellar black hole systems. The most direct method is to measure the *parallax* of the system. A parallax measurement is made by determining the tiny shift in observed position that occurs because the vantage point from Earth changes as the Earth orbits the Sun. The definition of a parsec, the basic unit of astronomical distance, is that at this distance the parallax shift is equal to one second of arc (hence *par-sec*). The importance of the parallax method to astronomy is demonstrated by its definition as the standard unit of distance; parallax measurements are the gold standard on which all other astronomical distances are based. Unfortunately, it is quite difficult to measure tiny fractions of an arcsecond, so most astronomical objects beyond a few hundred parsecs of distance, including most X-ray binaries, cannot be measured in this way. However, radio parallaxes have recently been used to determine the distances of a few X-ray binaries. Combining radio observations from many stations across the Earth can lead to positions accurate to better than a milliarcsecond, which corresponds to a distance of a kiloparsec. The distances of several of the closest X-ray binaries have been measured in this way.

But parallax measurements cannot be made for most X-ray binaries. Nevertheless, some of their distances are moderately secure. In transient systems, the accurately known orbital parameters (period, mass function, mass ratio, and inclination) lead to a reliable distance determination. Information about the size of the Roche lobe filled by the companion star is combined with the temperature determined by its spectral type. Again, the Stefan-Boltzmann relation is used, this time to calculate the luminosity of the companion star. The distance can be determined by comparing the luminosity with the observed flux. This distance is subject to all the uncertainties associated with the orbital parameters and also requires accounting for any interstellar absorption along the line of sight. The calculation of the distance is in most cases the limiting factor in measuring the spin of the black hole from continuum emissions.

Key requirements for this method are that the accretion disk extends to the ISCO and that observed emission is dominated by thermal radiation from the inner disk. Equivalently, if the accretion flow is in the form of a standard α-disk, this is likely to be true but is certainly not always true. In X-ray binaries, these measurements can be made when the inner disk is a significant component of the flux, typically in the soft X-ray state. In AGN, the situation is even more difficult, since the thermal emission from the inner disk is typically in the far ultraviolet, a wavelength range in which photons do not propagate through the galaxy. Therefore, this method has been used only for X-ray binaries. Even then, questions abound: Does the disk really extend all the way to the ISCO? Are the black hole mass and the distance well measured? Does the material inside the ISCO contribute significantly to the emission? One hopeful sign for this method is that in some sources, the size of the ISCO has been measured many times with the same result, even though the X-ray flux and spectrum were quite different. This consistency strongly suggests that the potential problems arising from an understanding of the accretion flow itself are not significant: however, the binary parameters must still be accurately determined.

8.4 Consequences of Spin for Jets and Other Phenomena

A great deal of effort has gone into trying to measure the spin of black holes. One might ask why. Mass determinations are crucial to establishing that an object is indeed a black hole, and in understanding how black holes are made. But the spin measurements assume that the system being observed contains a black hole and so cannot be used to confirm the nature of the system.

One possible consequence of black hole spin is the creation and collimation of the relativistic jets observed in many AGN and some X-ray binaries. That jets in these two kinds of systems are similar (up to the scaling by mass of the black hole) suggests that some deep physics associated with accretion onto black holes must be invoked to account for them. One of the earliest models was that of Blandford and Znajek, who suggested that frame dragging from a rapidly spinning black hole could create the necessary jet power.[1] This model would imply that the power in the jet should be at least loosely correlated with the spin of the black hole. An alternative mechanism, now more widely accepted, proposed that the jets are collimated by winding up the magnetic field of the disk, in which case the spin of the black hole is much less important. Recent results from the thermal disk method appear to favor the magnetic model: while the X-ray binary with the strongest observed jet does indeed appear to have a spin parameter close to unity, another source which is believed to have a strong jet has the lowest observed spin parameter ($a < 0.5$ at high confidence, and consistent with zero). Taken at face value, this result would likely rule out the Blandford-Znajek mechanism, but it should be noted that the jet in this system was observed only indirectly and that the usual caveats about the accuracy of the spin measurements apply. It seems likely that more numerous and more accurate measurements of the spin of black holes will provide a definitive ruling on whether the Blandford-Znajek mechanism is in fact responsible for the production of the observed superluminal jets.

Another situation in which the spin of the black hole is important is in the details of colliding or merging black

[1] R. Blandford and R. Znajek, 1977, *Monthly Notices of the Royal Astronomical Society*, 179: 433.

holes. When the event horizons of two black holes come
into contact, the black holes rapidly merge. Energy carried
away in the form of gravitational waves is potentially
observable, as described in the next chapter. The merger
process will proceed differently depending on the spin—
and the orientation of the spin—of the black holes, and
so the expected behavior at the moment of merger, when
most of the gravitational waves are produced, depends
critically on the spin of the individual black holes. The
spin of the merger product is the vector sum of the orbital
angular momentum of the orbit and that of the two
original black holes, less whatever angular momentum is
shed in the merger. Supermassive black holes may be built
up by gas accretion or by successive mergers of black
holes, whose spins should be randomly aligned. If the
primary growth mechanism is many small mergers, the
spins should cancel out, whereas if the growth is due to
only a few mergers, or by mergers with very lopsided mass
ratios, or by gas accretion, a black hole could have a very
significant spin. Thus observations of spin in black holes
are closely connected with the merger process that results
in supermassive black holes.

In general terms, spinning black holes raise issues that
are not present in standard Schwarzschild black holes. In
particular, the region of space between the ergosphere and
the event horizon can produce a variety of very strange
phenomena, some of which are potentially observable.
If black holes are used to test and explore the more
extreme predictions of general relativity, examples with
well-determined nonzero spin seem likely to produce the
most interesting results.

9

DETECTING BLACK HOLES THROUGH GRAVITATIONAL WAVES

The vast majority of our information concerning the cosmos arrives in the form of photons (electromagnetic radiation). Each photon can be characterized by four numbers, namely, the energy and arrival time of the photon, and two numbers characterizing its position on the sky.[1] So, our information about the Universe can be imagined as being contained in a very long table, each entry of which consists of four numbers, as well as the accuracy of each determination.

Viewed in this light, the enormous progress in observational astrophysics over the past half-century can be viewed as advances in the range and resolution of our determination of these numbers. The range in energy detectable has been of particular importance. A century ago we could observe only optical photons. The radio, X-ray, infrared, ultraviolet, microwave, and gamma-ray ranges were successively opened, so that now observatories

---

[1] In fact, additional information is carried by the polarization of the radiation, which in some cases is extremely useful. But special observational apparatus is required to detect polarization, and most celestial radiation is unpolarized.

can detect essentially all photon energies capable of propagating through interstellar and intergalactic space. Each new wavelength regime revealed a qualitatively different Universe. At the same time, new technologies dramatically improved the temporal, spatial, and spectral precision of our measurements of photons.

While we can expect further improvements in our ability to detect and measure electromagnetic radiation, it is possible that the next great advances in observational astrophysics will come from the detection of other kinds of information altogether. A number of such nonelectromagnetic "messengers" are known. Cosmic rays, which are charged particles (electrons, protons, alpha particles, and others), have been studied for some time. They are difficult to analyze, however, because the trajectories of charged particles are changed by the presence of magnetic fields, so the specific origin of a particular cosmic ray is often hard to determine. Nevertheless, observations of the intensity and energy of cosmic rays has led to insight into high-energy cosmic events. Neutrinos from celestial events have also been observed, most notably from the Sun and from the supernova that was observed in 1987 in the Large Magellanic Cloud. Neutrino astrophysics is also an active field, but the very low detectability of neutrinos seems likely to restrict observations to a few objects that have the right combination of intensity and proximity.

Currently, there is great excitement about the possibility of directly detecting an entirely new "celestial messenger," namely, *gravitational radiation*. The existence of gravitational waves is a prediction of general relativity, and current technology has put us very close to being able to detect them directly. The strongest sources of gravitational radiation are expected to be merging black holes. Since we expect such mergers to occur, both between stellar-mass and supermassive black holes, the detection of gravitational radiation would provide a new way not only to explore gravitational physics but also to look for and to study celestial black holes.

9.1 Gravitational Waves and Their Effects

General relativity holds that the presence of mass warps the space and time around it. When a mass moves, the space-time distortion moves with it. If the mass changes its motion (that is to say, accelerates), this creates a wiggle in the space-time distortion that propagates away from the mass at the speed of light. This wiggle takes the form of a gravitational wave. The generation and propagation of gravitational waves is in some ways analogous to that of electromagnetic waves, which are created by the acceleration of a charged particle. However, in the case of an electromagnetic wave, quantum mechanics describes how the wave can also be analyzed as a particle (a photon). In the absence of a good theory of quantum gravity, consideration of "gravitons" is more problematic.

One common form of acceleration occurs when the masses in question are in orbit around each other. For circular orbits, there is a constant amount of acceleration in a smoothly changing direction, which in turn produces a steady flow of gravitational waves (see figure 9.1). These waves carry energy with them; this energy is extracted

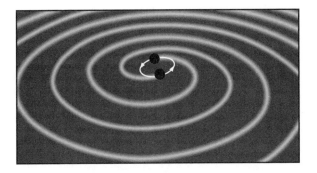

Figure 9.1. Gravitational waves propagating outward from black holes in orbit.

from the binary orbit. Thus the existence of gravitational waves is an intrinsically non-Newtonian process, in that it reduces the total energy of the orbit, causing the two objects to spiral inward rather than to remain in a permanent stable orbit with a constant period, as Newtonian theory requires. The orbital period will decrease due to gravitational radiation as

$$\dot{P}/P = -3.68 \times 10^{-6} \frac{M_1 M_2}{(M_1 + M_2)^{1/3}} P^{-8/3},$$

where M_1 and M_2 are the masses of the binary components measured in solar masses, and the orbital period P is measured in seconds. The form of the mass term on the right-hand side means that systems in which the masses are roughly equal decay more quickly, while the high negative power of P means that binary stars with orbits of a few days or more do not exhibit measurable effects. Another consequence of this equation is that as the period decreases, the rate of the period change increases, creating a runaway condition—that is, an ever-decreasing orbital period that decreases more and more quickly. By integrating the change in P over time, one can compute how long it will take for the binary period to formally approach zero. This in-spiral time T_0 represents a maximum time after which the two orbiting objects will merge and can be written

$$T_0 = 2.33 \times 10^{-3} \frac{(M_1 + M_2)^{1/3}}{M_1 M_2} P^{8/3} \text{years}.$$

Orbiting objects much larger than their Schwarzschild radius will merge sooner, since the binary separation will still be large when their surfaces touch. Other ways to extract energy from a binary orbit most notably include tidal effects, which can become very large for objects whose radius is comparable to the orbital separation. But for orbiting black holes T_0 is a good measure of the time until merger. For a binary system containing two objects of 5 $M_\odot$ in an orbital period of 10 hours—comparable to many observed X-ray binaries—the merger time is only about 10 million years, which is much less than the age of the Universe. As we will see, there are observed systems whose parameters are such that we expect such mergers to take place.

The luminosity of the gravitational wave radiation emitted by a black hole merger event is immense. There is a "natural" gravitational wave luminosity that can be constructed by combining the fundamental constants G and c in a form with units of luminosity, namely, $L_0 = c^5/G = 3.6 \times 10^{59} \text{ erg s}^{-1}$, which is 26 orders of magnitude greater than that of the Sun. This value is greater than the electromagnetic luminosity of the entire Universe. L_0 is roughly the luminosity of the gravitational wave radiation emitted by a system in which the velocities are strongly relativistic ($V \approx c$), and the spatial scales are comparable to the Schwarzschild radius. Thus L_0 is an upper limit to the gravitational wave radiation emitted by any particular event, since c and R_s limit the velocity and size of any physical object, and the luminosity falls off quite rapidly as V becomes less than c, and R becomes greater than R_s. However, as a black hole binary system merges, the velocity and scale of the two objects approach R_s and c, so just before the merger the luminosity in gravitational waves becomes very large.

It is interesting to note that this maximum luminosity is not dependent on the mass of the objects, just on their velocity and the size relative to the Schwarzschild radius. So, merging supermassive black holes approach the same limiting luminosity as merging stellar-mass black holes. At first this seems counterintuitive—surely the more massive systems should emit more energy than the lower-mass systems. But it must be remembered that the timescales for the more massive systems are greater than those for the lower-mass systems. The orbital period just before merger scales with the Schwarzschild radii and thus with the mass of the black holes. Thus more massive systems emit at high L for longer than the lower-mass systems do, so the total energy emitted is correspondingly greater, even though the instantaneous maximum luminosity is the same.

By the same token, the frequency of the gravitational radiation depends on the mass of the merging objects.

The frequency of the radiation when L approaches its maximum is related to the inverse of the orbital period at the last stable orbit. Since the orbital velocity is the same fraction of the speed of light at this point, the period is linearly proportional to the circumference of the orbit and thus to R_s and to the mass of the black hole. For merging stellar-mass black holes, frequencies are in the kilohertz range. Similar phenomena are expected, for supermassive black holes but with frequencies at or below a millihertz. Thus the best range for seeking these kinds of events is several orders of magnitude on each side of 1 Hz.

9.2 Binary Pulsars

While gravitational waves have not yet been observed directly, one class of systems in which their consequences can be readily observed is the *binary pulsars*. Pulsars are magnetized spinning neutron stars. When the magnetic pole of the pulsar points toward the observer (once per rotation), a "pulse" of radiation is observed. These pulses have been observed and timed to great accuracy by radio observatories since the late 1960s, and more recently in other wavelengths as well. The short periods of the pulsars (typically a few seconds or less) indicate that the rotational periods and thus the size of the objects themselves must be small, much smaller than the categories of stars that were known at the time. Indeed, pulsars were the first direct evidence for the existence of neutron stars.

In the 1970s, several binary pulsars were discovered whose changing velocity curves determined from the

Doppler shifts of the pulsation period clearly indicated that the pulsar was in a binary orbit of a few tens of hours with another neutron star. The tiny radius of the neutron stars means that tidal effects are negligible, so the relativistic effects on the orbits could readily be seen. In particular, the decay of the orbit due to gravitational radiation is clearly apparent in several systems. The discoverers of the first binary pulsar, Russell Hulse and Joseph Taylor, were awarded the Nobel Prize in physics for this thrilling demonstration of what had previously been a theoretical concept.[2]

Given the high precision of pulsar measurements, a number of relativistic effects could be observed, including the precession of the orbital periastron (whose presence in the orbit of Mercury was one of the first empirical confirmations of general relativity), the gravitational redshift due to the changing distance between the two sources, and the expected decrease in the orbital period due to gravitational radiation. These observations completely constrained the parameters of the orbit and the mass of the orbiting objects, which could be determined to great precision, and precisely predicted second-order relativistic effects such as the *Shapiro time delay*. These second-order effects have since been observed and are measured to be consistent with the predictions of general relativity. These observations of the orbital motion of the binary pulsars now provide severe constraints on possible non-Einsteinian

---

[2]Hulse and Taylor's accounts of their discovery, recounted in their Nobel Prize lectures (which can be found on the Nobel Prize site http://www.nobelprizes.org), provide a wonderful description of the interplay between technical expertise, deep physical insight, and good fortune in physics.

gravitational theories. But while these systems provide very powerful indirect evidence for the existence of gravitational wave radiation, a yet more stringent test would be the direct detection of gravitational waves, in the manner that electromagnetic radiation is detected by devices ranging from the human eye to photographic film to digitized silicon devices.

9.3 Direct Detection of Gravity Waves

Attempts to detect gravitational waves directly have a long and somewhat checkered history. The basic approach is to observe small changes in space-time caused by the passage of a gravitational wave by observing the minute resulting changes in size of an object as the wave passes through it, which requires extraordinarily precise size measurements. For most expected astrophysical sources of gravitational waves, the precision must be $\Delta L / L \approx 10^{-22}$ or better: that is, the expected change in size or separation will be only 10^{-22} of the size or separation of the objects being measured. Given that the size scale of an atomic nucleus is a few femtometers (1 fm $= 10^{-15}$ m) it is remarkable that such experiments can even be contemplated—for a 1 m object, the measurement precision would have to be better than a millionth of a single nucleus!

The first attempts at detecting gravitational waves used large bars of aluminum. The effect of gravitational waves at the resonance frequency of the bars would be amplified, thus allowing measurements of a change in length. Starting in the late 1960s, Joseph Weber claimed positive results using this method. However, the results could not be duplicated in other laboratories, and in any case the claimed amplitude of the observed effects was much larger than the expected size of the astrophysical events thought to cause them. These results are therefore generally discounted.

Elaborate efforts are currently under way to detect gravitational waves using the technique of *laser interferometry*, in which light emitted by a laser is combined with light from that same laser that has traveled down a different path. The basic design is that the laser emission is sent down two different paths at right angles, then reflected and recombined.[3] Depending on the precise distances of the paths, the peaks and troughs of the light waves either combine constructively, and the light is amplified, or destructively, in which case the light is dimmed. If the length of one of the arms changes even slightly due to a gravitational wave passing through the apparatus, the strength of the recombined signal will change significantly. Changes of length of a small fraction of a wavelength of the light can be detected by monitoring the strength of the combined light beam.

The frequency range of the gravitational waves observable from such an interferometer is related to the inverse of the travel time of the beam. So, for deca- and kilohertz frequencies, as expected from merging stellar mass black holes, the light travel time should be a fraction of a second, and thus the length of the light path would ideally be hundreds or thousands of kilometers. For mergers of

[3]This is basically the same arrangement as the famous Michelson-Morley experiment, which demonstrated that the speed of light is the same regardless of the motion of the light source and detector.

SMBHs, timescales of hundreds or thousands of seconds would be desirable, and thus beam lengths of millions of kilometers would be required.

This approach is now being implemented in the Laser Interferometer Gravitational Wave Observatory (LIGO), which is an experiment being carried out at two sites, one in Washington state and one in Louisiana. At both sites, a laser shines in two directions down a several-kilometers-long track to thermally and vibrationally isolated mirrors (see figure 9.2). Tiny changes in length in the two directions are recorded. Having two widely separated sites provides confirmation that any observed signal is celestial in origin, and any time delay between the effects at the two sites provides information on the direction to the source of the radiation, since gravity waves must propagate at the speed of light.

LIGO is a very delicate experiment, and a large team of scientists is working to reduce all the sources of noise in the system to levels where expected astrophysical signals could be detected. Key limitations on the precision of the measurements include seismic vibrations, thermal vibrations in the mirror and laser apparatus, and the ability to measure tiny changes in the strength of the combined light beam. LIGO is currently sensitive to displacements near 10^{-18} m. Given the kilometer size of the experiment, sensitivity is thus near $\Delta L / L = 10^{-21}$, tantalizingly close to gravitational wave amplitudes expected from astronomical sources, but no detections have yet been confirmed. Work is currently underway on Advanced LIGO, which should improve the sensitivity by more than a factor of 10, allowing frequent detections.

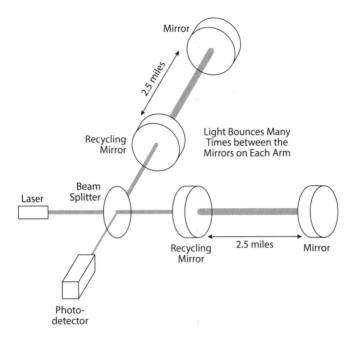

Figure 9.2. Schematic of LIGO setup. Laser light is sent down two long corridors and reflected to the start. When recombined, the interference patterns can reveal changes in the relative length traveled of a small fraction of the wavelength of the light. Extra precision is obtained by repeated reflections using the recycling mirror. After diagram from LIGO website (www.ligo.caltech.edu).

Ground-based interferometers are limited by the vibrations from the Earth due to human-caused seismic activity. These limitations apply particularly strongly to frequencies of less than 1 Hz, so low-frequency gravitational waves, like those from merging supermassive black holes, are unlikely ever to be detectable from ground-based experiments. This

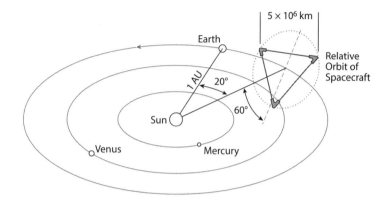

Figure 9.3. Planned orbits of the three LISA satellites. After Figure 5 from P. Aufmuth and K. Danzmann, 2005, *New Journal of Physics*, 7: 202.

drawback has prompted interest in creating a space-based gravitational wave observatory. In particular, a project called LISA (the Laser Interferometer Space Antenna) has been developed by a collaboration between NASA and the European Space Agency and is currently being reviewed prior to authorization (see figure 9.3). LISA will fly three spacecraft in an equilateral triangle several million kilometers on a side. Inside each spacecraft will be a free-flying, and thus vibrationally shielded, set of mirrors and a laser. Each satellite will constantly monitor distances to both of the others. The large distances and careful shielding of the mirror and laser apparatuses from all noise sources should allow for great precision and sensitivity at the relatively low frequencies associated with mergers of supermassive black holes. Some technology development is still required before this mission can be flown, but the primary limitation now appears to be financial—if full

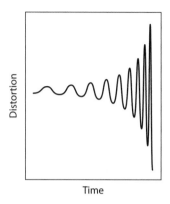

Time

Figure 9.4. Intensity as a function of time for a *chirp*. The characteristic of a chirp is that both the frequency and the amplitude of the signal increase exponentially with time.

funding were available, estimates are that LISA could fly within a decade or so.

9.4 Detecting Astrophysical Signals

As the binary spirals together ever more rapidly the gravitational radiation takes the form of a *chirp*, which is the term used for an oscillating signal whose frequency and amplitude both increase exponentially (see figure 9.4). Just before the merger, the gravitational waves reach a maximum strength, with a period equal to the orbital period just prior to the merger. In the case of neutron stars, which have a radius only somewhat bigger than their Schwarzschild radius, this frequency is of order $\approx 10^3$ Hz, corresponding to an orbital period of about a millisecond.

While these high-frequency gravitational waves occur for only a fraction of a second, the amplitude is so high that such an event would be observable by a device like Advanced LIGO over hundreds of megaparsecs. Our own galaxy is known to contain a few binary pulsars, each of which has a decay timescale of a few million years, suggesting that such an event will be generated every million years or so. But since the events may be observable in millions of galaxies, we would expect to see one at least once per year. Thus known systems will necessarily generate observable gravitational wave events quite regularly. It is this expectation that justifies the construction of the current generation of gravitational wave observatories.

While neutron star mergers generate strong gravitational radiation signatures, mergers of black holes are even stronger. Two black holes in orbit release gravitational radiation, and the chirp continues to rise in both frequency and amplitude until the event horizons are in contact. At that point the event horizons merge, creating a single black hole, and a last intense burst of gravitational wave radiation carries off energy and angular momentum until the merger remnant settles into the configuration of a single Kerr black hole. It seems likely that stellar-mass systems result in mergers between black holes or between black holes and neutron stars, in addition to neutron star systems like the binary pulsars. But the clear existence of double neutron star systems that must evolve to mergers demonstrates that events with frequencies of $\approx 10^3$ Hz almost certainly occur.

But supermassive black holes can also merge—and indeed they must if current ideas about the evolution of galaxies and the SMBH at their center are correct. As noted in chapter 6, SMBHs can grow through repeated mergers of smaller black holes. These mergers result in huge bursts of gravitational radiation, but with much lower frequencies than mergers of stellar-mass objects. The frequency of gravitational waves from a black hole merger are comparable to the orbital frequency near the event horizon. Observations of such events will be a direct test of hierarchical galaxy formation models and thus important to cosmology as well as to relativistic astrophysics. But as discussed earlier while these frequencies are accessible from planned space missions like LISA, they are unlikely to be observed from the ground.

Observation of any kind of black hole merger event requires that the gravitational merger signal—the chirp— be extracted from data containing sources of noise much larger than the signal itself. Fortunately, the repetitive nature of the wave signal as it passes over the detector will allow a very faint signal to be extracted, and sophisticated timing analysis allows signals with a known time signature to be identified with high precision. In the early stages of the merger, a classic chirp is expected, whose functional form is easy to define. However when the two event horizons approach each other, the pattern becomes more complex. But this is the moment when the signal is strongest, so it is of great importance to be able to accurately predict the expected pattern of the incoming radiation, which can be done by careful computational modeling of the space-time associated with the merger. The field of *numerical relativity* has advanced greatly in recent years, allowing ever more accurate computations of the expected wave patterns. These patterns depend on the initial configuration of the merging black holes—their orbit, their relative masses, and the strength and direction of the black hole spins. Thus libraries of expected patterns are being computed for comparison with the data that will eventually be obtained.

10

BLACK HOLE EXOTICA

Black holes are a staple of science fiction, where their effects are used as plot devices in a variety of ways. But the effects evoked by science fiction writers are generally not those observed by astronomers in the ways that are described in this book. Rather, they are extrapolations from the predicted behavior of black holes that have not yet been observed. But in principle these exotic behaviors are observable, and as science fiction correctly points out, they could have dramatic effects on people's lives under the right circumstances. In this final chapter we explore some of these predicted effects and the possibility that they might someday be explored in fact as well as in fiction.

10.1 Hawking Radiation

Relativity is in some senses a very classical theory. It is deterministic; that is, if the initial conditions of a system are precisely known, its further behavior can in principle be precisely predicted. It is also continuous, in that the theory remains intact if one extrapolates distances to zero. Relativity shares these characteristics with Newtonian theory, but the situation is very different in quantum mechanics. It is these differences that make a theory of quantum gravity so difficult (thus far impossible) to construct. However, one situation in which the interaction between quantum mechanics and relativity has been explored with some success involves *Hawking radiation*, a process through which black holes are expected to emit energy and ultimately evaporate.

Such a process runs counter to our basic ideas about black holes. Black holes, after all, are supposed to involve one-way processes, in which an object gains mass and energy but cannot lose them or emit anything. But space-time as derived from a metric is a continuous construct and by its nature does not allow for quantum effects. Nevertheless, those quantum effects must occur, particularly in the immediate vicinity of an event horizon. The event horizon around a Schwarzschild black hole, for example, is an infinitely thin spherical surface. But "infinitely thin" is a phrase that leaps out as demonstrating the need for a deeper theory. In particular, phenomena that occur so close to an event horizon that quantum uncertainties extend across the horizon clearly cannot be adequately described by relativity theory alone.

One important quantum phenomenon is the production of *virtual pairs* in a vacuum. In this process, a particle and its antiparticle are spontaneously produced but recombine quickly enough to conserve mass and energy within the quantum uncertainties. Predictions of particle interactions carried out using Feynman diagrams and other analytic tools require the creation of virtual pairs to explain a variety of observed phenomena. Hawking radiation arises when a virtual pair is created in the near vicinity of an event horizon. If one of the newly created particles crosses the event horizon, it cannot reemerge to recombine with its partner. The remaining particle can then travel away from the event horizon. When viewed from a distance, the black hole appears to be emitting a stream of particles. These particles carry off mass and energy from the black hole, which eventually evaporates.

One way to think about Hawking radiation is through the simple approximation that a black hole emits blackbody radiation with the peak wavelength of order the size of the Schwarzschild radius. This heuristic approach leads to results that reveal the basic features of Hawking radiation, and their implications for the observability and effects of this phenomenon.

The wavelength of maximum emission of a blackbody can be determined by differentiating the blackbody spectrum as a function of wavelength λ and setting the result equal to zero. The result is an expression for λ_{max} as a function of blackbody temperature T: $\lambda_{max} = hc/CkT$, where C is a numerical constant defined by $Ce^C/(e^C - 1) = 5$, or $C = 4.965....$ In dimensional units

$$\lambda_{max} T \approx 2.9 \times 10^{-3}$$

where λ_{max} is measured in meters and T in kelvins. Setting λ_{max} equal to the Schwarzschild radius than gives $T \approx 2.9 \times 10^{-3} c^2/2GM$ in standard mks units.

Given this temperature, we can determine the blackbody luminosity of the black hole using the standard formula $L = \sigma 4\pi R^2 T^4$, where the relevant radius is again the Schwarzschild radius. Since $R_s \propto M$, and $T \propto 1/M$, we find $L \propto M^{-2}$. Thus low-mass black holes will radiate far more than massive black holes.

The energy that generates this luminosity is extracted from the mass of the black hole, which can be converted into energy using the familiar expression $E = Mc^2$. Thus Hawking radiation reduces the mass of the black hole, which makes it radiate more intensely. So a low-mass black hole will undergo a runaway and proceed rapidly to complete evaporation. Since luminosity is defined by $L = dE/dt$, the time to evaporation t_{evap} will be

$$t_{\text{evap}} = \int_{Mc^2}^{0} (1/L)dE \propto M^3.$$

More sophisticated calculations than the simple model presented here show that

$$t_{\text{evap}} = \frac{5120\pi G^2}{\hbar c^4} M^3,$$

or in dimensional units

$$t_{\text{evap}} = 2 \times 10^{67} (M/M_{\odot})^3 \text{years},$$

where $M_{\odot}$ is the mass of the Sun. Thus stellar-mass black holes will take far longer to evaporate than the age of the Universe (currently estimated to be 1.4×10^{10} years), and SMBHs will take orders of magnitude longer than that. These enormous timescales show how tiny the effects of Hawking radiation would be on black holes of the kinds that we know exist—indeed, the accretion of a single optical photon every few years would counteract the effects of Hawking radiation for a stellar-mass black hole altogether. Thus Hawking radiation remains a theoretical prediction, not yet observable in any known situation.

We can also calculate the mass for which a black hole would evaporate over the current age of the universe, which proves to be $M_{evap} \approx 2 \times 10^{11}$ kg, or much less than stellar-mass or supermassive black holes. Since the evaporation time scales with a high power of M, most black holes with $M < M_{evap}$ will have evaporated, while black holes with $M > M_{evap}$ will still exist.

Another consequence of the strong dependence of Hawking radiation on M is that the final stages of black hole evaporation will proceed very rapidly. Thus one might expect to see a burst of radiation associated with the evaporation of a black hole, and if a class of primordial black holes with $M \approx M_{evap}$ existed, one might expect to see occasional bursts of radiation associated with the demise of these.

10.2 Primordial Black Holes

The preceding formulas show that there is a sharp cutoff between black holes with a mass of less than $\approx 10^{11}$ kg, which would evaporate over the course of the universe, and those with a higher mass, which would still exist. The strong dependence on mass means that black holes for which Hawking radiation is significant will radiate at increasing rates and proceed exponentially toward complete evaporation, while black holes a few orders of magnitude more massive will show little effect over cosmic time. Therefore the effects of Hawking radiation are relevant only for black holes with $M < 10^{11}\,\text{kg} \approx 10^{-19}M_\odot$. Standard evolution of stars and galaxies cannot lead to black holes this small, since pressure forces and chemical bonds will more than suffice to hold up objects of the required low mass against gravity. However, it is possible that such small black holes could have been created in the early Universe.

In its earliest stages, the Universe was much denser and hotter than it is now. Tiny fluctuations might then have given rise to objects that collapsed inside their event horizon. The density and temperature of the plasma is so great that ordinary matter is replaced by more exotic particles and antiparticles, which may not be able to hold such objects up against collapse. A number of specific moments in cosmic history have been suggested when such primordial black holes might have been created, and what masses they might have. Such hypotheses are hard to test observationally, since low-mass black holes will have long since evaporated, and isolated high-mass black holes are hard to detect.

However, black holes with evaporation times comparable to the age of the Universe might be readily observable. If such black holes were created in the early Universe, they might be completing their evaporation right now. Since Hawking radiation is a runaway effect, increasing in intensity as the black hole becomes less massive, an evaporation event will end in a sudden large increase in radiation, whose intensity might briefly be very high and thus observable. This effect was one of the more exotic explanations of the sudden γ-ray burst events that are observed throughout the Universe, but there is now much more compelling evidence that these are associated with particular kinds of collapsing stars.[1] Currently, astrophysicists are taking a different approach, using the absence of observable Hawking radiation events to set limits on the number of primordial black holes that could have formed in the early Universe, which in turn provides information on physical processes associated with the first moments of the Big Bang.

Some limits can be set on the formation of black holes at other masses. The radiation generated by the evaporation of low-mass black holes at much earlier times in cosmic history might have affected observable phenomena like the cosmic microwave background. No such effects are observed, and that limits the number of primordial black holes that could have been evaporating at some stages of the evolution of the Universe. Gravitational microlensing observations limit the number of more massive black holes which currently exist. It is clear that in most mass ranges there are too few primordial black holes to account for the observed dark matter. Nevertheless, direct clues to the possible creation, existence, and evaporation of primordial black holes are hard to come by, and significant populations of these black holes could be hidden in the Universe.

One possible way to explore the creation of primordial black holes is through particle physics experiments that reproduce (very briefly) the conditions that were present in the early Universe. The possibility that black holes might be produced in particle accelerators has received considerable press attention—given the popular impression of black holes as objects that can swallow up everything

---

[1] See J. Bloom, *What Are Gamma-Ray Bursts?* (Princeton, NJ: Princeton University Press, 2011).

nearby, one can understand why there might be some concern! Fortunately however, this concern is misplaced. Events as energetic as those in any particle accelerator occur frequently at the top of the atmosphere, when energetic cosmic rays encounter atoms in the ionosphere. This process has been going on for billions of years, and obviously the Earth has not yet been swallowed by a black hole. Furthermore, any black hole that could be formed would be of very low mass and thus would have high Hawking radiation. Very small black holes would evaporate essentially immediately and be transformed into streams of particles similar to those that created them. For example, a 1 kg black hole would evaporate in less than 10^{-16} second. Furthermore, any black hole in which the Hawking radiation is greater than the Eddington limit would not be able to accrete mass. Since the Hawking radiation of a low-mass black hole is high, while the Eddington limit is low, low-mass black holes of the kind that might be created in particle experiments cannot accrete under any circumstances.

10.3 Wormholes

One of the most enticing possible effects associated with black holes is that they might form *wormholes* through which widely separated parts of the Universe can be closely connected. This is perhaps the most common use of black holes in science fiction, since it provides an escape from the central difficulty of space opera, namely, that that light travel times between most celestial objects are vastly greater

than a human lifetime.[2] The basic idea is that two singularities at different points of space-time might be joined
by a "bridge" or "throat" which takes a different path
through space-time than would ordinarily be followed. If
this connection is shorter than the ordinary space-time
trajectory, then mass and energy might appear to travel
from one black hole to the other at faster-than-light speeds.
Alternatively, it is possible that the connection might allow
changes in the time coordinate, creating a time machine.

To use the wormhole as a practical means of transporta-
tion, matter and energy would have to be able to enter one
of the black holes and emerge from the other. This would
require that the singularity not be surrounded by an event
horizon but, rather, that it be a "naked" singularity. In
principle, such naked singularities can exist, but in general,
they are not stable. To hold the black hole open requires
matter with negative energy density, for which there is no
current evidence. Even if such exotic matter did exist, there
is no conceivable sequence of events that would lead to the
creation of a wormhole. Penrose has proposed the "cosmic
censorship" theory, which essentially says that singularities
are not permitted to be naked. Extensive computational
simulations of the gravitational collapse of various configurations of ordinary and exotic matter have produced
ambiguous outcomes, but no convincing demonstration of
a real situation in which a naked singularity would result.

---

[2]The invention of the concept of wormholes provides an especially intimate
connection between science fiction and science fact. In his book *Black Holes
and Time Warps: Einstein's Outrageous Legacy* (New York: Norton, 1994), Kip
Thorne provides a charming account of how some of the key ideas behind
wormholes were developed at the behest of Carl Sagan for use in Sagan's novel
Contact.

If a wormhole did exist, what might it look like? Here the science fiction representations (e.g., *Star Trek*) depart completely from reality. The manifestation of a wormhole would presumably be that interstellar matter accreted by one black hole would appear to emanate from another. Thus one end of the wormhole would be a "white hole" in which matter and energy was being extruded. This would look quite different from a black hole emitting Hawking radiation, since the rate at which mass/energy was emitted would depend on how much was being accreted at the far end of the wormhole, rather than the mass of the black hole itself. Given the wide range of radiating astrophysical objects, it would be difficult to establish that a particular observed object was in fact a wormhole. Many other seemingly much more plausible explanations would need to be eliminated. It might be even harder to identify the location of the other end of the wormhole, since if these objects exist at all, there is no reason for their two ends to have any particular relationship to each other. Nevertheless, the possibility of wormholes being discovered and perhaps being technologically exploited by a sufficiently advanced civilization remains, even though there is currently no theoretical demonstration that they can form and no empirical evidence that suggests that they exist.

10.4 Multiverses

One final suggestion that might be contemplated is that a separate universe, perhaps with natural laws different from our own, might exist inside the event horizon of a black hole. This is one version of the *multiverse* concept, in which a variety of universes with a variety of characteristics exist, of which our Universe is only a particular example. Much of the attraction of this concept lies in how it explains the peculiar friendliness of our Universe to complexity and thus to life. As noted in chapter 5, most values of the total energy of the Universe result in either a single black hole or a dispersed lifeless void, but our Universe lives in the infinitesimally small range that allows for many individual objects to exist. Many other natural laws and values of constants also appear to require such fine tuning for life to exist. One way to address these seeming coincidences is to postulate that many universes with all sorts of laws exist, and that most of them are in fact quite simple, but that we necessarily exist in a Universe that allows complexity.

One version of this idea that has a particularly close link to black holes is described by Lee Smolin in his book *The Life of the Cosmos*.[3] Smolin postulates that the complexity and fine tuning we see in our physical universe is created in much the same way that complexity arises in biological systems, namely, through natural selection. If a new universe is birthed within each black hole, then a universe that creates lots of black holes will create more numerous progeny and thus be "fitter" in the evolutionary sense than a universe that produces zero black holes or one black hole. If each child universe has characteristics similar to its parent, then universes that produce the largest possible number of black holes will soon be the dominant

---

[3] L. Smolin (Oxford: Oxford University Press, 1997).

species. Smolin suggests that this may be testable: if one simulates a large number of possible universes, it should turn out that ones that actually exist are among the ones that produce the most black holes.

There are some problems with this formulation. In particular, it isn't clear what rules would be used to generate the small but nonzero changes between parent and child universes that are required for the evolution of the multiverse. Nor is it clear that our understanding of cosmic processes is sufficient to accurately determine how many black holes a universe with a particular set of characteristics may have. Nevertheless, it is fascinating to imagine that the existence of a universe conducive to life might be due to the existence of black holes, and that at least one example of the resulting life-forms might begin to explore and understand how that happened.

GLOSSARY

Accretion Disk (2.2, 2.7):   A thin rotating disk of gas surrounding a compact object. Viscous forces within the disk lead to gradual accretion of the gas onto the compact object.

Accretion Disk Corona (4.3):   Hot optically thin gas above an accretion disk. It generates significant X-ray emission in some X-ray states of X-ray binaries.

Active Galactic Nuclei (chapter 5):   Accreting supermassive black holes in the center of a galaxy.

ADAF (2.3):   Advection Dominated Accretion Flow. One kind of radiatively inefficient accretion flow, in which most of the energy in the accretion flow is advected onto the central object. The flow is generally more radial than in an accretion disk.

AGN (see Active Galactic Nuclei)

α-Disk (2.7):   A simple accretion disk model originally developed by Shakura and Sunyaev in which the viscosity is parameterized by a constant α multiplied by the height of the accretion disk and the sound speed.

Anthropic Principle (6.0):   The idea that human existence necessitates that the Universe satisfy very particular conditions.

Binary Pulsar (9.2):   A rotating magnetized neutron star in a binary orbit with another neutron star. Binary pulsars show clear but indirect evidence for gravitational wave radiation and are used as tests of general relativity.

Blazar (5.3):   AGN in which a relativistic jet is directed toward the observe. The Doppler boosting of the jet results in unusually bright, energetic, and variable jet emission.

Bondi-Hoyle Accretion (2.1):   Spherically symmetric accretion. It is generally not realized in physical situations, since angular momentum is generally significant in accretion flows.

Boundary Layer (4.6):   Material accreted onto the surface of a white dwarf or neutron star. Radiation from a boundary layer should not be observed from an accreting black hole.

Bremsstrahlung (2.5):   Radiation emitted when electrons are deflected by the electromagnetic charge of some other particles.

Cataclysmic Variable (4.4):   A generic term applied to an accreting white dwarf.

Chandrasekhar Limit (4.4, 4.8):   The maximum mass of a white dwarf. If the Chandrasekhar limit is exceeded, the star cannot be supported by Fermi pressure of electrons.

Chirp (9.4):   A signal whose frequency and amplitude increase exponentially with time. The gravitational waves emitted by merging black holes are expected to be in the form of a chirp.

Compton Scattering (2.5):   Emission mechanism due to interaction of electrons and photons, resulting in dramatic changes to the observed photon spectrum.

Eddington Limit (2.1):   An upper limit on the accretion rate imposed by radiation pressure from the accretion luminosity.

Ellipsoidal Variation (4.5):   An orbital change in the flux observed from the companion star of an X-ray binary due to the nonspherical shape of the star.

Equation of State (4.8):   The relationship between density and pressure in a gas.

Ergosphere (1.2, 8):   The region between the event horizon and the Schwarzschild radius of a rotating black hole.

Escape Velocity (1.1):   The speed required to escape the gravitational attraction of a spherical object: $V_{esc} = \sqrt{2GM/R}$.

Event Horizon (1.1):   The boundary beyond which information and radiation cannot travel to a distant observer. For a nonrotating black hole, the event horizon occurs at the Schwarzschild radius.

Fermi Pressure (4.8):   Pressure that arises because two particles (fermions) are prohibited from occupying the same volume simultaneously.

Globular Cluster (6.1.3):   A dense cluster of stars. X-ray binaries are unusually prevalent in globular clusters, presumably due to dynamic interaction between compact objects and binaries and other stars. Globular clusters are sometimes thought to harbor intermediate mass black holes.

Gravitational Redshift (1.2, 8.2):   The change in wavelength of an observed photon relative to its rest wavelength: $z = 1/\sqrt{1 = R_s/r} - 1$.

Gravitational Microlensing (4.7):   The bending of light rays from a visible star by an unseen star in the line of sight, resulting in a magnification of the visible star. Used to detect isolated stellar-mass black holes.

Gravitational Wave Radiation (chapter 9):   Radiation emitted by the acceleration of masses.

Hard Binary (6.1.3):   A binary star whose binding energy is greater than the kinetic energy of an incoming third object.

Interactions between hard binaries and other objects tend to harden the binary further.

Hawking Radiation (10.1):   Emission from a black hole due to quantum mechanical effects.

Hydrostatic Equilibrium (4.8):   The balance of gravitational forces inward with pressure forces outward.

Innermost Stable Circular Orbit (ISCO) (7.1):   The closest closed orbit to a black hole. The distance from the event horizon depends on the spin of the black hole, and thus the location of the ISCO is used to measure black hole spin.

Intermediate-Mass Black Hole (chapter 7):   A black hole more massive than a star, but less massive than a supermassive black hole. The existence of IMBHs is disputed.

Laser Interferometry (9.3):   A method for very precise distance measurements used in attempts to directly observe gravitational waves.

Line Emission (2.5):   Emitted radiation that is concentrated in a small range of wavelengths.

M-σ Relation (5.5):   An observed correlation between the mass of a supermassive black hole at the center of galaxy and the mass of the stars in the central portion of the galaxy.

Magnetic Braking (6.1.2):   A mechanism for loss of angular momentum from a rotating or orbiting body, in which the angular momentum is transferred to corotating material through magnetic fields.

Magnetohydrodynamics (3.2):   The study of fluid flows in magnetic fields.

Magnetic Reynolds Number (3.2):   A nondimensional number reflecting the importance of fluid flow compared with magnetic dissipation: $R_m = VL/\eta$.

Mass Function (4.5):   A quantity with units of mass computable from the observed radial velocity curve of binary stars. The mass of the unobserved component of the binary system must be greater than or equal to the mass function. The large value of mass function of some X-ray binaries provides evidence that the compact objects cannot be neutron stars and are most likely black holes.

Mass Segregation (6.1.3):   The tendency for more massive bodies to fall to the center of a star cluster over a relaxation time or longer.

Metric (1.2):   A line element that defines the separation between space-time events. The Schwarzschild metric is used for non-rotating black holes, and the Kerr metric for rotating black holes.

Multiverse (10.4):   The concept that there may be multiple universes, each of which may have different laws of physics.

Neutron Star (4.4):   A star made primarily of neutrons, supported by Fermi pressure of the neutrons.

Numerical Relativity (9.4).   Techniques for computer simulations of relativistic situations; actively pursued in exploring the expected signals from gravitational waves from black hole merger events.

Parallax (8.3):   A method of determining distance based on the slight changes in an object's apparent location in the sky due to the motion of the Earth around the Sun.

Penrose Process (1.2, 8.4):   The process by which energy can be extracted from the spin of a black hole.

Quasar (5.1):   A very bright AGN: sometimes used to denote bright AGN detected in radio wavelengths.

QSO (5.1): Quasi-Stellar Object. A very bright AGN sometimes used to denote bright AGN detected in optical wavelengths.

Radiative Transfer (2.6): The process of transmission of radiation through gas or plasma, which may affect the observed spectrum of the radiation.

Radio Galaxies (5.2): A class of galaxies that emit unusually large amounts of radio emission.

Relaxation Time (5.5): The time required for a cluster of objects to exchange significant amounts of energy through gravitational interactions.

Reverberation Mapping (5.5): A method of determining the mass of an SMBH by comparing the delay between luminosity changes in the accretion disk and the broad-line region with the luminosity of the AGN.

Schwarzschild Radius (1.1): The radius of the event horizon of a nonrotating black hole: $R_s = 2GM/c^2$.

Singularity (1.1): The point or ring at the center of a black hole where the density of matter is formally infinite.

Sgr A* (5.4): The radio source at the center of the Milky Way galaxy. Measurements of the motions of stars near Sgr A* have demonstrated that the galactic center harbors a black hole with a mass about 3 million times that of the Sun.

SMBH (chapter 5): Supermassive Black Hole. A black hole with a mass of $10^6 \, M_\odot$ or greater generally found in the center of AGN and other large galaxies.

Superluminal Jets (3.1): Jets emitted by AGN (and occasionally X-ray binaries) that appear to propagate faster than the speed of light. This is an optical illusion created by the effects of special relativity.

Supernova (6.1): An explosion caused by the collapse of the center of a massive or accreting star.

Synchrotron Emission (2.5): Radiation emitted due to acceleration of electrons by magnetic fields.

***Uhuru* Satellite (4.1):** The first orbiting X-ray observatory, launched in 1970.

Ultraluminous X-Ray Binaries (ULXs) (7.1): Extragalactic X-ray binaries that radiate at significantly greater than the Eddington limit for an object of a few tens of solar mass.

Unification (5.2): The idea that many different observable phenomena in AGN are due to the effects of orientation to the observer.

White Dwarf (4.4): A star held up by Fermi pressure of the electrons. The final stage of stellar evolution for stars with a final mass of less than the Chandrasekhar limit.

Wormhole (10.3): A pair of black holes joined by several different paths through space-time. To be stable, the throat of the wormhole must be surrounded by matter with negative energy density. There is currently no empirical evidence for wormholes or for matter with negative energy density.

Velocity Curve (4.5): A plot of radial velocity against time for an orbiting star. The period and amplitude of the velocity curve help determine the mass of the compact objects in X-ray binaries.

X-Ray Binaries (4.2): Binary stars in which accretion onto a compact object generates large numbers of observable X-rays. High-mass and low-mass X-ray binaries refer to the mass of the mass-losing star relative to that of the accreting compact object.

X-Ray Burst (4.6): A brief X-ray flare observed in X-ray binaries that occurs explosively in a fraction of a second. It is thought to arise from thermonuclear ignition of material accreted onto the surface of a neutron star. X-ray bursts should therefore not be observed from black holes, and indeed they are not.

X-Ray States (4.3):   Different configurations of spectra and timing characteristics of X-ray binaries.

INDEX